Evolution of Mathematical Concepts

E. B. Tylor, 1832-1917, English anthropologist, pioneer student of the evolution of culture. (Culver Pictures.)

Evolution of Mathematical Concepts

An Elementary Study

Raymond L. Wilder

Professor Emeritus, University of Michigan

JOHN WILEY & SONS, INC.

New York   London   Sydney   Toronto

By the same author

Introduction to the Foundations of Mathematics, Second Edition, 1965

QA
9
W57

Library of Congress Catalog Card Number: 68-28508
SBN 471 94414 9
Printed in the United States of America

To Una

Preface

ADMITTEDLY, I think, mathematics is one of the most important cultural components of every modern society. Its influence on other cultural elements has been so fundamental and widespread as to warrant the statement that our "most modern" ways of life would hardly have been possible without mathematics. Appeal to such obvious examples as electricity, radio, television, computing machines, and space travel to substantiate this statement is unnecessary; the elementary art of calculating is evidence enough. Imagine trying to get through the day without using numbers in some fashion or other!

But it is not the importance of mathematics that motivates this study; rather it is the desire to determine, if possible, how and why mathematical concepts, such as *number* and *geometry,* were created and developed. We know quite a bit about how the individual mathematician creates and develops his concepts; studies on the psychological level have been made, and eminent mathematicians, notably Poincaré and Hadamard, have contributed from their own experiences. But these form only a part of the story. No mathematician operates in a vacuum; his interests are determined not only by the state of mathematics in his time, but by his contacts with mathematical colleagues throughout the world. The late anthropologist Ralph Linton stated hypothetically that ". . . if Einstein had been born into a primitive tribe which was unable to count beyond three, life-long application to mathematics probably would not have carried him beyond the development of a decimal system based on fingers and toes." Since it took the combined efforts of several "civilizations" and innumerable mathematicians over a period of four or five thousand years to produce a decimal system, it is questionable if even the genius of an Einstein could have done so in a single lifetime. Einstein (and the same holds for other mathematical geniuses) was able to accomplish what he did because of a combination of factors, only one of which was his unquestioned genius. And most of these factors are of a cultural nature.

Mathematicians themselves seem prone to ignore or to forget the cultural nature of their work and to become imbued with the feeling that the concepts with which they deal possess a "reality" outside the cultural milieu—in a sort of Platonic world of ideals. Indeed, some mathematicians seem to be completely lacking in the insight that the modern physicist has attained—the recognition that even his observations, as well as his concepts, are colored by the observer. How much more must this be the case in a subject like mathematics, where the conceptual has gradually gained primacy over the observable?

No one expects to find *laws,* couched in the logic and language of a given culture, to which the physical universe strictly conforms. The physicist formulates so-called laws purely as a mode of rationalizing the environment and predicting its behavior; he does not assert that nature *obeys* those "laws." Insofar as mathematics was grounded in the real physical environment, it sought out laws—of arithmetic, geometry, and the like—to which the universe conformed. But once mathematics became a mature cultural element, it appeared to proceed on its own, as though oblivious of "reality."

However, mathematics did not develop independently of cultural forces, some peculiar to its own nature, any more than did physics, art, or other cultural components. Attempts to change the direction of mathematical research by individuals who deem the tendencies prevailing at a given time to be "wrong," seem to be of little avail. Only strong environmental and internal pressures, such as are sometimes imposed by war, dislocation forced by political changes, radical alterations in the host culture, "crises" in mathematics itself, and the like, appear to be effective in changing the course of mathematical development. As witness we have the stagnation of ancient and medieval Chinese mathematics, reflecting the static character of the host culture; the decline of Greek mathematics, whose cause is perennially debated, but in any case was part of a general cultural decline (internal and environmental); and the effect on American mathematics—and on mathematics in general—of the influx of European mathematicians to the United States prior to World War II. In recent times stimulation of new research, and an unusual enhancement of the status of its practitioners, has been effected in mathematics much as in the more obvious case of physics—primarily because of the political environment.

Moreover, I believe that both mathematics and philosophy of mathematics stand to gain from an investigation of the evolution of mathematics. If it be true that "In darkness dwells the people which knows its annals not,"* it is equally true that the mathematician who ignores the evolutionary forces that have shaped his thinking, thereby loses a valuable perspective. To know *history* is not enough; dates, biographical materials, and the like are important, but they form only part of the "artifact" collection for a study of this kind. Moreover, now that the evolution of culture has become a recognized theory in anthropology, a study of this kind should

---

* Inscription on facade of William L. Clements Library, University of Michigan (due to the late Professor Ulrich B. Philips).

be of general interest, particularly to anyone who has an interest in the genesis of the ideas that form his cultural environment.

The chief obstacles in such an undertaking are the inadequacies of early records. It is a moot question whether it is easier to study the processes of culture in general or those of a particular cultural item such as mathematics. In studying culture as a whole there is a huge store of artifacts from which to draw conclusions, whereas for a particular cultural item, such as mathematics, the available material is usually relatively limited. On the other hand, limitation serves to cut down the number of complexities and pinpoints attention. One is reminded of the study of the evolution of the horse. Much as the study of the evolution of a particular form of life can suggest patterns for more general forms, so can a study of a particular cultural item, such as mathematics, have significance for the general forms that cultural evolution takes.

Recognizing, however, that mathematics is notoriously technical, it seems advisable to focus principally on the elementary aspects of number and geometry. Their evolution exhibits essentially all the characteristics that are found in the development of more advanced parts of mathematics. Not only did mathematics begin with number, but number concepts cut through every field of mathematics in some form. And it is the part of mathematics of which every civilized person (and noncivilized for that matter) must cultivate some knowledge in order to cope with his environment, both social and physical. It should, therefore, not be too technical for general comprehension. Only in Chapter 3 are there any technicalities, but these have, I hope, been presented in such a way as to be assimilable by the nonmathematician. The reader who does not wish to read all of this chapter will find sufficient material in the remainder of the book to understand the general import, I believe.

What I have tried to do here is to study the mathematical subculture from the standpoint of an anthropologist, rather than that of a mathematician. Of course, since I am a mathematician, this approach is subject to the same hazards that are encountered by a social scientist who essays to study his own culture; it is difficult to get outside one's own culture and examine it dispassionately. But mathematics is so profoundly technical that it would be well-nigh impossible for one outside the profession to pierce the curtain of symbolically clothed and highly abstract concepts that envelop it. The analogue of the "informant" system, ordinarily used to obtain knowledge of the customs and beliefs of a primitive culture, would not be found workable in the case of mathematics.

I wish to disclaim any attempt at philosophizing. I have intended purely an inquiry into the development and behavior of a certain part of culture. It is impossible to avoid philosophical concepts completely; after all, mathematical philosophy has influenced the development of number, especially during Greek times. If, on the other hand, certain of my conclusions seem to have only a philosophical basis, it must be considered as due to the sinuous boundaries that separate what we call philosophy of science and scientific theory. I can make this clearer, perhaps, by using the analogy of religion, which has been studied purely from an anthropological viewpoint with little or no reference to the philosophy of religion, except as this forms part of the content of a particular religion.

The professional mathematician should not expect to find in this book a rigorous treatment of such topics as the real number system, since this is not intended as a textbook. This is a book *about* mathematics—mathematics as a cultural phenomenon—and not a contribution to mathematics per se. If the professional mathematician who reads it should gain a better insight into the nature of his work, he will have gotten all I can hope to offer him. On the other hand, I trust that the nonmathematical reader, especially the student of sociology and anthropology, will find a real understanding of what mathematics is all about, and that he will ignore those minor technicalities that he may encounter here and there and will read on to the end.

For the sake of the nonmathematical reader, I have included an Introduction concerning the nature of modern mathematics, not only from the standpoint of what it is and how it developed, but also from the standpoint of what is happening in the teaching of mathematics. For the sake of both the mathematical and non-mathematical reader, I have commenced the text proper with a chapter (Preliminary Notions) that, I hope, contains sufficient material concerning cultural anthropology, and the intrinsic nature of the decimal system that we use daily, to take care of any technicalities not elucidated thereafter. All of this material may be omitted, and reading commenced with Chapter 1, if the reader is already familiar with its substance.

References to the Bibliography are by name of author and date; thus "Bell, 1931, p. 20" refers to the item opposite 1931 under Bell, and specifically to page 20 thereof. Cross-references to items within the book are generally made by citing chapter and section; however, citations to sections within the same chapter omit the chapter number.

I am indebted to too many colleagues and students to attempt a complete citation of the names of those who have contributed to the formation of my ideas. I am particularly indebted to my anthropologist friend, Professor Leslie A. White, as well as to my children Betty Ann Dillingham (who read and criticized the first draft) and David, both social scientists, for the benefit of many discussions of related materials; none can be held responsible for ideas expressed herein. To my colleague Professor Phillip S. Jones and my former student Professor Alice B. Dickinson, I owe much helpful discussion and encouragement. Special thanks are due to Mrs. Mary Ann Sober for her extraordinarily efficient and willing secretarial assistance. I should acknowledge, too, my indebtedness to the universities and colleges that have afforded me opportunity to lecture on mathematics from a cultural point of view. It was because of the encouragement of members of my audiences that I decided to write this book. I hope that those who heard my lectures and who happen also to read this book will recognize at least a faint resemblance between the two. Thanks are also due to the Michigan Institute of Science and Technology and to the Florida State University for their aid, through the media of research professorships during the academic years 1960-1961 and 1961-1962, respectively, in providing some of the time necessary to the study and research involved in an undertaking of this kind.

R. L. Wilder

Ann Arbor, Michigan
May 1968

Contents

xviii

Evolution of Mathematical Concepts

Introduction

1   CONCEPTIONS OF THE NATURE OF MATHEMATICS
Although every civilized person uses mathematics to some extent, if
only to add up his ready cash, and is indirectly affected by it in
almost all phases of his life, it is probably true that no subject is so
much misunderstood. By this I mean that the *nature* of the subject
is not generally understood; few can be expected to be acquainted
with its technicalities, since not only are these extraordinarily
complex, but few elect mathematics as a profession.

Even though we start our mathematical training as soon as
we learn to talk and, in civilized countries, continue it through
the elementary school and frequently through the high school and
into the university, our opinions regarding the nature of mathematics and its relations to other aspects of culture show an
enormous variety. Moreover, this phenomenon does not exclude
the opinions of professional mathematicians. Consider the following examples:

"In the pure mathematics, we contemplate divine truths,
which existed in the divine mind before the morning stars sang
together, and which will continue to exist there, when the last of
their radiant host shall have fallen from heaven" (Bell, 1931, p.
20). This came from the lips of the famed Edward Everett, whose
oration, in the opinion of contemporaries, eclipsed Lincoln's
revered address at Gettysburg.

1

"Mathematics is a tool which ideally permits mediocre minds to solve complicated problems expeditiously"—from a textbook on physics (Firestone, 1939).

"I believe that mathematical reality lies outside us, and that our function is to discover or *observe* it, and that the theorems which we prove, and which we describe grandiloquently as our 'creations' are simply our notes of our observations"—from a statement made by a noted modern mathematician (Hardy, 1941, pp. 63-64).

". . . We have overcome the notion that mathematical truths have an existence independent and apart from our own minds. It is even strange to us that such a notion could ever have existed" —a joint statement made by a well-known modern mathematician and equally well-known science writer (Kasner and Newman, 1940, p. 359).

". . . It is the merest truism, evident at once to unsophisticated observation, that mathematics is a human invention"—from the pen of a distinguished modern physicist (Bridgman, 1927).

These opinions variously contain elements of mysticism, practical utility, Platonism, and "common sense." Apparently none of them is designed as a *definition* of mathematics, but rather as an opinion on the *nature* of the subject. Presumably no one but a professional mathematician or a philosopher of science would essay a definition of mathematics. However, anyone is entitled to state his opinions regarding the nature of mathematics and what he considers to be its role.

Mathematics has grown so much during the past fifty years, while also altering the conceptions inherited from our forebears, that a mathematician trained during the first two decades of this century would be hopelessly out of date if he had not kept up with the changes. I am reminded of a remark by a pathologist who worked in the Great Lakes area; while discussing the effects on the thyroid of the general use of iodized salt in that region, he stated that a pathologist of 1920 would not know how to interpret what he might see in thyroid tissue today. Similarly, a mathematician who might have been thoroughly up to date in 1920 would today be unable to comprehend current journal articles if he had not kept up with advances in his field. Consequently, what might have been a correct description of the nature of mathematics in 1920 would likely be very inadequate for present-day mathematics.

These considerations suggest a question: Are the changes all to the good? Or has the subject taken a "wrong turning," to use the phrase with which some have characterized the ancient Greek developments in mathematics?

2 MATHEMATICS IN THE SCHOOLS   Another question relates to the pedagogical aspect of mathematics: How are the changes affecting the teaching of mathematics?

The university courses on the graduate level have generally kept up with the changes; the situation could hardly be otherwise when the professors who teach these courses have been the ones responsible, through their research, for the occurrence of the changes. On the undergraduate level, the effects have also been noticeable, especially in the major universities. However, the medieval type of mathematics still taught in the secondary schools is only now being challenged by the experimental texts being written under the auspices of the National Science Foundation and other agencies. This means, of course, that parents are beginning to feel the effects of the changes and are commencing to wonder to what "newfangled" notions their children are being subjected. So long as the changes affected only the graduate students in the universities, the mathematical community could go its way in the proverbial "ivory tower" manner. But when Mary and Johnny begin coming home with mathematics texts written in a language that Mom and Dad cannot understand (even though they may both have bachelor's degrees from good universities), something strange must be taking place, and perhaps the school board should look into it!

A fundamental difficulty is that the parent often does not

understand that mathematics is not something that was handed down by divine revelation to some mathematical "Moses" in bygone times. Mathematics is something that man himself creates, and the type of mathematics he works out is just as much a function of the cultural demands of the time as any of his other adaptive mechanisms. Nearly every primitive tribe invented number words to some extent; but it was only when ancient civilizations such as the Sumerian-Babylonian, the Chinese, and the Mayan developed trade, architecture, taxes, and other "civilized" appurtenances that number systems were invented. Of the pre-Hellenic types of mathematics, the Sumerian-Babylonian was the most advanced. In fact, it was so far advanced (as we have only recently learned) that one wonders if the parents of the little Babylonian Mary and Johnny perhaps called on the temple scribes to justify some of the ideas they were teaching.

Perhaps no other subject is so susceptible to such extremes of good and bad teaching as is mathematics, and much "bad" teaching stems from a failure to bring out the excitement of creating mathematics. And a *sine qua non* for making mathematics exciting to a pupil is for the teacher to be excited about it himself; if he is not, no amount of pedagogical training will make up for the defect.

One difficulty, no doubt, is that mathematics requires such refined and complicated symbolic techniques for its expression. The teacher who becomes so utterly engrossed in mastering techniques for operating with symbols that he forgets their conceptual background is fated to lose the interest of his pupils and will also leave them with some of the misconceptions of mathematics that are so current today. On the other hand, as our discussion of the evolution of mathematics should make clear, even the elementary concept of number could not advance very far until a suitable symbolic apparatus—a numeral system—was set up.

A good case can be made for the thesis that man can be distinguished from other animals by the way in which he uses symbols (*cf.* White, 1949, Chapter 11, for instance). Man possesses what we might call *symbolic initiative;* that is, he can assign symbols to stand for objects or ideas, set up relationships between them, and operate with them on a conceptual level. So far as has been ascertained, no other animal has this faculty, although many animals do exhibit what we might call *symbolic reflex* behavior. Thus a dog can be taught to lie down at the command "Lie down"; and to Pavlov's dogs, the bells signified food. In a popular magazine a few years ago, a psychologist was portrayed teaching pigeons to procure food by pressing certain combinations of colored buttons. These are examples of symbolic reflex behavior—the animals do not create the symbols, but can learn to react to them just as they react to other environmental stimulants.

As an aspect of our culture that depends so exclusively on symbols, as well as on the investigation of relationships between them, mathematics is perhaps the furthest from comprehension by the nonhuman animal. However, much of our mathematical behavior that was originally of the *symbolic initiative* type drops to the *symbolic reflex* level. We memorize multiplication tables and then learn special devices (called algorithms) for multiplying and dividing numbers. We memorize simple rules for operating with fractions and formulas for solving equations. These are justifiable laborsaving devices, and the professional mathematician often puts much effort into devising them. However, the professional mathematician understands the purpose of what he is doing, while the pupil who learns only the devices usually does not even comprehend why they work. Processes whose understanding demands symbolic initiative have been placed on the symbolic reflex level.

As a result, a considerable amount of what passes for "good" teaching in mathematics has become of the symbolic reflex type, involving no use of symbolic initiative. This is the drill type of teaching, which may enable stupid John to get a required credit in mathematics but bores the creative-minded William to such an extent that he comes to loathe the subject. What essential difference is there between teaching a human animal to use an algorithm to find the square root of a number and teaching a pigeon to punch certain combinations of colored buttons that will produce food? Perhaps emphasis on the symbolic reflex type of teaching is to some extent justified when the pupil is very young—closer to the so-called animal stage of his development. But as he approaches maturity, certainly more emphasis should be placed on his symbolic initiative.

3 HUMANISTIC ASPECTS OF MATHEMATICS   During the course of its evolution, mathematics has developed so many humanistic aspects that for its devotees it may be said to play the role of a humanity. In order that this statement not be misunderstood, I should explain that I am using the term "humanity" to indicate an aesthetic pursuit, usually employing the tools of art, literature, or music—in a nutshell, the use of symbols of some sort to achieve beauty, simplicity, harmony, or other types of what are generally considered to be aesthetically satisfying qualities.

There is evidence that in this sense of the term even the Babylonian mathematics began to assume humanistic features; to put this another way, the Babylonian mathematicians seem to have indulged in a little "mathematics for its own sake." If one were asked to select a portion of mathematics that would likely prove of little use in the nonmathematical affairs of life, one would probably choose number theory. It is a part of mathematics that an amateur can understand to considerable extent, since it involves only the "natural numbers"—that is, the numbers 1, 2, 3, . . . , with which we count. Evidently the Babylonians made a start in it.

Like mathematics today, Babylonian mathematics was a science—what I prefer to call a number science (see Chapter 1, Section 3.2), since it consisted only of the natural numbers and their relations, with extensions to sexagesimal fractions and rules for applications to weights and measures. Perhaps I should explain that I think of "science" as the construction of conceptual theories or models, of physical or other (e.g., social) realities, with a view to adaptation and prediction, and possibly involving descriptive and empirical activities designed to furnish bases for and checks on the theories. At first, Babylonian mathematics was not a science, any more than a few number words constitute a science in a primitive culture. But eventually the Babylonians developed the concept of number, which was an achievement of no mean order. What we call a *scientific concept* had evolved, and this was important because once it was born, one could go on to imagine larger and larger numbers than one had ever noted in the physical world and, furthermore, study their *properties*. And having found such properties, one could use them to *predict,* for example, just how many bricks a mason would need to build a wall. For one now knew how to combine numbers by addition and multiplication and how to embody them in rules for obtaining needed information concerning trading and building, for instance. Of course we take all of this for granted today, but to the Babylonians it was apparently an exciting experience while it was evolving. And in the process they had evidently begun to discover that their numbers had certain qualities that evinced features of harmony and simplicity—humanistic features. They probably did not go very far in the investigation of these qualities, although research in Assyriology may yet show that they did.

The Pythagorean school of the sixth century B.C. in Greece (Magna Graecia) went much further (see Chapter 1, Section 3.4). Much of their terminology, for instance "amicable," "friendly," and "perfect" numbers, had humanistic overtones, They became so entranced with the seemingly inexhaustible variety of marvelous features that the natural numbers displayed and with the new varieties of applications that they found for them (as in music) that they ultimately attributed to these numbers a mystical character and gave them a prominent place in their philosophy. I think it is no exaggeration to say that the Pythagorean mathematics (which, we must not forget, also included geometry) was more a humanity than a science.

If we turn to the evolution of geometry, we find the same trend from the purely scientific to humanistic. Babylonian mathematics contained little that we would call geometry—only a set of rules for mensuration of no more significance than rules for finding the interest on a sum of money. But in Greece, abstraction from the patterns employed by the mason, carpenter, and surveyor led to the notions of triangle, rectangle, polygon, the regular solids, and the like, finally evolving into a well-constructed theory based on a few simple axioms and employing the methods of deductive logic. It was now a real science, since the theory seemed to be well representative of patterns perceptible in the physical world. But it also exhibited humanistic traits that played a strong part in its development. Perhaps the reader will have heard how the Pythagoreans finally discovered that one can find a pair of line segments such that no matter how small a unit of length is selected, it will not measure off the length of *both* line segments exactly; if it measures one exactly, it will not measure the other. Such pairs of segments are called *incommensurable*. For example, the side of a square and the diagonal of the same square are incommensurable. Since the Pythagorean geometry had assumed commensurability of all line segments as part of its foundation, a "crisis" developed that could be resolved only by providing a new foundation admitting incommensurability. At the same time, the paradoxes of Zeno concerning the infinity of points on a straight line segment struck at the soundness of the scientific aspects of mathematics.

Now, I am certain that the average Greek was blissfully unaware of this crisis, just as the modern American was unaware of an analogous situation in the foundations of mathematics that developed around the turn of this century. What use has the carpenter, engineer, or even physicist for such a subtle affair as the distinction between commensurable and incommensurable line segments? They know that measurements of length are only approximate—that one can never arrive at an exact mathematical measure for a physical object—so why bother with such matters?

Fortunately the Greek philosophers—and in those days mathematicians were philosophers—could not be satisfied with imperfection; it marred the beauty and distorted the simplicity of the geometrical edifice. So they set out to reconstruct geometrical theory and came up with solutions that have had a profound effect on all subsequent mathematics and the sciences that use it as a tool. For one thing, the deductive method, which to the Greeks had a strong aesthetic appeal, was adopted by other sciences, and today there are signs that even the social sciences will make use of the axiomatic method for theoretical developments. As for mathematics itself, the axiomatic method is today one of its most important research tools. Attention should be called, too, to the way the humanistic side of mathematics contributed here to its scientific side. The desire for perfection led to the development of a method for constructing theories, that is, the axiomatic method, without which modern mathematics and science could hardly get along.

Let us consider another example of the same type of phenomenon. This concerns the so-called parallel axiom. It can be stated in many different forms; perhaps the simplest is the assertion: if L is a line and p is a point not on L, then there is only one line through p that is parallel to L and lies in the same plane with L. Euclid did not state the axiom exactly this way, but what he said was equivalent to it. And there are signs that he was not satisfied with the axiom, in the sense that he suspected it might not be necessary. If one is stating some fundamental assumptions—the axioms—on which one intends to build a theory, he usually prefers that no axiom be a logical consequence of the others. The technical term for this is "independence"; we prefer that all our axioms be *independent*. Thus if we take some statement in a theory that can be deduced from the axioms and deliberately include it among the axioms, the axioms are no longer independent. There would be nothing *wrong* with doing this—except that one feels that it is *aesthetically* bad to do so. The only time we ever do this sort of thing in mathematics is in those instances where we use a system of axioms as a basis for teaching a subject; then we sometimes take a theorem that is very difficult to prove and place it among the axioms.

To come back to the Greeks, they evidently felt that there was something aesthetically unsatisfactory about using the parallel axiom. This feeling continued through later centuries, it being generally suspected that the parallel axiom could be deduced from the other axioms of geometry as stated in Euclid's *Elements*.

The attempts to show that the parallel axiom could be deduced from the other axioms ultimately took the form of trying to show that if the *denial* of it were added to the other axioms, contradiction would result; that is, the classical *reductio ad absurdum* type of argument. The best known of these attempts was that of the Italian Saccheri, whose work in this regard was published in 1733 in a book whose Latin title (*Euclides ab omni naevo vindicatus*) has been translated as *Euclid Freed of Every Flaw*. One can infer from this title that Saccheri thought he had accomplished his purpose and had actually shown that denial of the parallel axiom leads to contradiction. However, this work is one of the best examples of the "law": "If the facts do not conform with the theory, they must be disposed of."[1] Had Saccheri not been so utterly convinced of the nonindependence of the parallel axiom, he would probably be credited today with the invention of the non-Euclidean geometry. But he could not bring himself to such heresy, so he "lamely forced into his development an unconvincing contradiction involving hazy notions about infinite elements" (Eves, 1953, p. 124).

It was inevitable that the "truth of the matter" would be ultimately discovered, and anyone familiar with the way in which the modern sciences evolve will not be surprised to learn that it

[1] Possibly due to Hegel; my own acquaintance with this "law" stems from a formulation by the psychologist N. R. F. Maier.

was discovered virtually simultaneously during the first third of the nineteenth century by at least three mathematicians, namely Gauss, Bolyai, and Lobachevski. Gauss was too sound a mathematician to commit Saccheri's error, but like Saccheri may have been so afraid to commit heresy that he never published his results. Bolyai and Lobachevski did, and their ideas were so much alike that the suspicions and charges of plagiarism that generally accompany such circumstances were given vent to.[2] The type of non-Euclidean geometry that they had worked out is now called hyperbolic geometry. Some twenty years later, Riemann completed the picture by inventing another type of non-Euclidean geometry, called elliptic geometry.

This is another case in which "aesthetics" furnished the chief motive for what must have seemed, to any sane and upstanding citizen, a monumental waste of time—like musical composition or poetry writing(!), even if not so accessible to appreciation by the "man in the street." It probably still seems so to the so-called "practical" person, provided that he ever hears about it. Yet there is a direct line from such a seemingly "impractical" matter as the question of the independence of an axiom in an ancient Greek geometry to its answer in the non-Euclidean geometries of Bolyai, Lobachevski, and Riemann, thence to the theory of relativity (in which Riemann's geometry was a basic tool), and ultimately to the development of nuclear fission. If ever an argument were needed to convince the dispensers of research grants of the "utility" of supporting basic research in pure mathematics—which usually means the "humanistic" elements in mathematics—one has such an argument here. Of course, one may feel that it was a disaster that nuclear fission was ever discovered. But one should

[2] For an interesting account of such cases, see Merton, 1957. The same author also makes out an excellent case for the thesis that multiple discovery is the rule, rather than the exception, in science; see his article, Merton, 1961.

not blame mathematics or science for their technological by-products any more than one should invest musical theory and practice with responsibility for rock 'n' roll.

Many more analogous instances could be given, in which the aesthetics of "mathematical humanism" helped lead the way. For example, according to folklore, an ancient Greek poet related that King Minos was not satisfied with the size of the tomb built for his son and ordered that the tomb be doubled in size. He stated that to achieve this each *dimension* of the tomb should be doubled. Recognizing that this was incorrect, the Greek geometers took up the problem: How can one double the volume of a cube while retaining its cubical form? Thus, according to folklore, arose the problem of the "duplication of the cube." According to some historians, Menaechmus (a tutor of Alexander the Great) invented the conic sections in order to solve the problem (*cf.* Eves, 1953, p. 83). [However, Neugebauer (1957, p. 226) conjectures their origin in the theory of sundials.]

Now, regardless of the true origin of the problem, probably any Greek artisan worth his salt could duplicate a cube—at least well enough for his purposes. But what the Greek mathematicians wanted was the aesthetically satisfying, mathematically *exact* solution. Here again was a pursuit of purely humanistic motivation; possibly the Greeks had no other use for conic sections. But if the Greeks had not invented them and worked out their properties, they would not have been available to the astronomer Kepler some 2000 years later as a medium for expressing his famous laws, and possibly there would be no law of gravitation and no "lunik" to photograph the other side of the moon.[3]

It is very difficult to impart to anyone who has not himself experienced the thrill of creating mathematics, how similar are the experiences one undergoes in, say, musical composition or the creation of an abstract painting and in the creation of mathematical

[3] For an interesting analysis of the significance of this and other examples, see Boyer, 1959.

concepts. Perhaps it is as well not to try. I am reminded of the time, years ago, when a noted mathematician, Emil Artin, lectured for the scientific fraternity Sigma Xi. Although Artin's major specialty was algebra, he was also a topologist. The nature and classification of knots and braids have long been of concern to topology, and Artin succeeded in completely classifying braids (knots are still unclassified). He devoted his lecture to a description of his work on braids. As he was a master of exposition, most of what he said was quite clear to his audiences (which consisted for the most part of nonmathematicians). But on one occasion someone rose to remark, "Your talk was most interesting. But of what use is such work anyway?" Artin's reply was, "I make my living at it!" Fortunately he realized the futility of defense.

It is not surprising that many professional mathematicians consider mathematics to be an art, for certainly creative work in mathematics does share many common features with such artistic pursuits as music and painting. Moreover, the inspiration for many advances in mathematics has come from the artistic impulses of their creators. But this is true also of creative work in other fields of science, especially as they build up sizable bodies of theory; theoretical physics is a good example. However, the *cultural* significance and meaning of a human activity are not determined primarily by the motives of its individual practitioners. More important and fundamental from a cultural point of view is the role that the activity plays within the culture in which it is imbedded. For example, to its devotees, the importance of a religion usually lies in the solace and emotional satisfaction it gives them; as one of the basic components of a culture, however, religion acts more as an instrument of conformity, unity, and cohesion within the culture. Similarly, although the humanistic aspects of mathematics may be more important from the standpoint of the individual mathematician, it is as a basic science that mathematics functions within our culture (as witness its support together with the other sciences by the National Science Foundation and other agencies).

I am inclined to think that this also contributes to the schism that sometimes threatens to develop between the so-called pure and applied mathematicians—a miniature reproduction of the schism that threatens now and then to develop between the "sciences" and "humanities" in the broader university community. This is an age of specialization, and this is just as true in mathematics as elsewhere. The day of the mathematical universalist who "knew all mathematics" is gone forever; the average mathematician can hope only to acquire as broad a base as he can in the time available, in order to be able to do his bit in advancing the frontiers of knowledge before he reaches senility. This has led not only to a division of mathematicians into algebraists, geometers, analysts, logicians, statisticians, and so on, but each of these categories has its subcategories. However, it is the broader division into "pure" and "applied" to which I now refer.

It is apparently impossible to define the term "applied mathematics." In the "good old days" it was not necessary, for ever since the term "mathematician" became a meaningful designation, until quite modern times, "pure" and "applied" mathematics were usually found together. This was, I think, a very healthy situation. Since mathematics has its roots in the physical and social environment, one found therein a source of new ideas. The Babylonian "number scientist" was dependent on his environment for ideas, but ultimately he began to derive inspiration from the properties that he discovered in the natural numbers. And so long as a mathematician works in this way, developing his theoretical base, but constantly keeping his eye on his environment in order to derive inspiration from the interplay of theory and the part of his environment that is modeled within the theory, I would call him an applied mathematician. But when he becomes so enthralled with the mathematical concepts that he confines his study to their properties without any regard for his environment, he becomes a "pure mathematician"—one to whom the humanistic aspects of mathematics are paramount.

In this connection the general picture of the evolution of mathematics is revealing. It shows that the tendency is to build concepts suggested by the environment and then to generalize these concepts so that a higher level of abstraction is reached. In the pure mathematical realm, these concepts seem to take on a life of their own, and the research mathematician sometimes acquires a feeling that he is being led by the concepts instead of the other way around! For example, Heinrich Hertz, the discoverer of wireless waves, said: "One cannot escape the feeling that these mathematical formulas have an independent existence and an intelligence of their own, that they are wiser than we are, wiser even than their discoverers, that we get more out of them than was originally put into them."[4] What has happened is that mathematics has added concepts of its own to the world of so-called reality, so that its domain of application includes not only the physical and social environment, but the cultural environment—the growing bulk of mathematical theory having itself become part of the cultural environment. Consequently, a point of view that, for the sake of disputation, divides mathematics into its purely scientific and its humanistic aspects, sets up an unnatural dichotomy. The two are inseparable.

4 MODERN "REFORMS" IN MATHEMATICAL EDUCATION

The opposition of pure to applied mathematics has probably had some influence in producing disagreements concerning the attempts to modernize the secondary school curriculum, already referred to. The attitude taken by the "reformers" seems to be that "bad" teaching is not necessarily the fault of the teacher, but is more than likely due to the lack of adequate up-to-date conceptual materials both in the teacher's background and in the material that he has been given to teach. It is not that one wants to replace algebra, geometry, and trigonometry with something new, but rather to give them the benefit of a modern approach that recognizes the im-

---

[4] Quoted by Bell (1937, p. 16).

portance of an intuitive understanding of the conceptual background. One of the chief critics of the "reforms" has conceded in
allegorical fashion that "Our teachers of mathematics have been
teaching carpentry instead of architecture and color mixing instead
of painting."[5]

The urge for modernization appears to have a cultural impetus, which leads one to conjecture that the objections to it are
doomed to be of little avail, for what is happening has all the ear-
marks of a cultural change that cannot be diverted from its course.
If by the new methods mathematics can be made easier to grasp
so that more can be covered in the elementary curriculum than has
been the case, it seems destined to happen. As already mentioned,
the colleges have been much more responsive to change. In this
connection some facts are of interest.[6]

At Williams College in 1843, plane and solid geometry and
what is now considered high school algebra were offered in the
freshman year; in the sophomore year, more Euclid and some
navigation, surveying, spherical trigonometry, and conic sections.
In the junior year mathematics was taught for only one third of
the year, consisting of astronomy and some "fluxions," the latter
being Newton's name for calculus. In the senior year no mathematics was offered. At Oberlin, the situation was the same except
that no fluxions were taught—so no calculus at all. At Princeton
"they were very enlightened: they called fluxions differential and
integral calculus. But they jammed it into half the junior year and
gave only astronomy to seniors."

Thus "It took most of the first two years of college to learn
what is now taught in high school. . . . One could not go beyond
calculus in college," and this "despite the fact that mathematics,
Greek, and Latin constituted almost all of college education." Not

---

[5] M. Kline, in the *New York University Alumni News,* October 1961.
[6] Obtained from an address by C. Stanley Ogilvy to the Hamilton
College chapter of Phi Beta Kappa, as published in the *Key Reporter,* Vol.
25 (1959-1960).

until 1876 when Johns Hopkins instituted graduate work in mathematics could one get any mathematics beyond calculus. Contrast this with the fact that today many of the preparatory academies are offering calculus, and that it is gradually forcing its way into the larger public high schools wherever there is staff competent to teach it.

These matters are evidence, of a kind that the general public can perceive, of the continuing evolution of mathematics. However, although they may be called a proper part of the evolution of mathematical education, they are not inherently in the direct line of the evolution of mathematics itself. For the latter, one must study mathematics and its history, observing how it evolves under the influence of forces that act both within itself and without. The simplest part of mathematics to study for such a purpose is the development of the concepts of number and geometry. Consequently, in this book the case histories cited will be from these domains, with emphasis on the domain of number—one of the most profound, basic, and useful of man's intellectual achievements.

Preliminary Notions

In order that the material in this book not be obscured by technicalities, this preliminary chapter presents a brief survey of (1) the elementary portions of the theory of culture as it is employed in anthropology and (2) the manner of representation of numbers according to the place value system (with varying bases). The reader who is already acquainted with these matters should proceed immediately to Chapter 1.

1   THE NOTION OF CULTURE   The word "culture" has a variety of uses, such as in agriculture (cultivation of the soil), biology (a developing microorganismal mass), and society (e.g., the "cultured" person). The use of the term here will, however, be exclusively in the anthropological sense; or perhaps we should say in *an* anthropological sense, since anthropologists give varying definitions of the term. (See, for instance, Kroeber and Kluckhohn, 1952). Probably the most common meaning for the term in anthropological literature is: a collection of customs, rituals, beliefs, tools, mores, and so on, called *cultural elements,* possessed by a group of people who are related by some associative factor (or factors) such as common membership in a primitive tribe, geographical contiguity, or common occupation.

    1.1   *A culture as an organic whole*   It is important to realize, in this definition, the significance of the phrase "related

by some associative factor"; for this implies that the cultural elements are not independent of one another, but form a collection
whose parts are related to one another and influenced by one
another in manifold, and frequently unnoticed, ways. Thus a collection of Chinese habits of eating, Pueblo pottery, and English
eating utensils would not form a "culture," although they would
constitute elements of the Chinese, Pueblo, and English cultures,
respectively. But these kinds of cultural elements, as elements of
the *same* culture (e.g., English), will usually be found to have
influenced each other's development and uses. On a grander scale,
one can cite the relation between the technological status (e.g.,
agricultural) of a primitive people and its religious beliefs and
rituals. As cultural elements of the same group of people, they
have a profound influence on one another.

Just how the elements of a culture influence one another may
not be at all obvious and usually is not of interest except to the
social scientist. That influential relationships do exist and make of
a culture a kind of organic whole is, however, easily demonstrated
in the cases of certain cultural elements where recognition of the
relationships forces itself into the open. For example, that the introduction of the automobile has had an influence on American
culture has been discussed over and over in the public press.

1.2 *Relations between cultures and peoples*   Least
understood, perhaps, and certainly the most controversial is the
relationship between the people possessing the culture, that is, the
cultural group, and the culture itself. To bring out the kinds of
problems involved, it should first be stated that there are at least
four types of entities here: (1) the particular group of people
whose culture is under consideration; (2) the individual persons
in the group; (3) the culture possessed by the group; and (4) the
cultural elements. In the cases of (1), (2), and (3), one can
usually be quite precise as to just what group, and so on, is involved. (Incidentally, notice that we have ignored here, for reasons
of brevity, the various types of subgroups that may occur on levels

between that of the entire group of people and that of its individuals.) In regard to (4), one must decide just what are to be considered as the elements that make up the culture. How one does this will usually depend on one's purposes. For example, in the culture of an urban United States community it may suffice for some purposes to lump all religious manifestations under one heading—the religious element of the culture. For other purposes, however, it may be necessary to go to a lower level and distinguish between Christian, Jewish, Buddhist, and other religions as cultural elements; or, conceivably, to a still lower level, distinguishing between the Catholic element, the Baptist element, and so on. In the same manner, all scientific activity may be lumped together and considered as one cultural element; or, for the particular problem in hand, it may be necessary to consider the academic aspects of science as one cultural element, while applied science as it is practiced in the industrial laboratories is considered as another cultural element.

The flexibility of these conventions is most striking, perhaps, in the use of the term "culture" itself. What is considered a cultural element in one context may be the total culture in another. We spoke previously of the culture of an urban community. However, if one is studying the culture of the United States as a whole, one may speak of the "urban element" as one of the cultural elements constituting the culture of the United States as a whole. The culture of a given urban community then becomes a particular example of such a cultural element.

Most difficult, unquestionably, are the relations between the pair (1), (2), and the pair (3), (4); specifically, between (1) and (3), between (2) and (3), between (1) and (4), and between (2) and (4). Regarding (1) and (3), one can ask, for instance, "Where did the people get their culture?" A partial answer, certainly, is that they inherited it from their forebears. In the case of language, for example, this is quite obvious, as it would also be in the case of a state religion. But cultures are not unchanging; even language undergoes certain transformations, as also does religion, though perhaps more slowly. Recall that we used the term "organic whole," and it is in the nature of organisms to undergo changes. In particular, the American culture of today is certainly different from that of 1900.

Let us focus for the moment on two cultural elements that have already been mentioned, namely language and the automobile. It is a curious fact, long known to linguists, that long-term changes in languages can be formulated as "laws." It seems to make no difference who or what kinds of people use the languages; the changes that the languages undergo over the centuries follow certain patterns—the "laws." On the other hand, there are plenty of examples in which individual words have been introduced into a language by individual persons, such as the word "normalcy," which was introduced by the late President Harding. A little analysis would show that Harding obtained this word from already existing words and word forms, just as the medical researcher may invent a term from known Latin or Greek words, which will ultimately find its way into common discourse ("arthritis," "bacteria," etc.). The linguist is very little interested, however, in how words may be introduced by individuals; his main interests lie in the way in which languages are constructed from their elements (phonemes, words, etc.) and in the ways they evolve.

The case of the automobile is interesting since the history of its evolution is generally well known and since it was not the invention of a single individual. Before the automobile could be invented, a veritable host of other cultural elements had to be present. In addition to the obvious mechanical know-hows and gadgets (gears, cutting tools, etc.), the availability of a suitable fuel and knowledge of its chemical composition were necessary; also, the economic need or receptiveness to make feasible the marketing of the invention. In short, one can make the assertion that the then-current state of the Western culture (now including

English, French, etc., as well as American) had to contain a complex of elements—mechanical, economic, chemical—and natural resources, as well as a variety of relations that had already been formed between these, before one could even *think* of the object that we now call automobile. Moreover, it is safe to surmise that any, or perhaps even all, of the so-called inventors of the automobile could have died in their infancies, and yet the automobile would have been invented. Similar remarks can be made about the airplane. Orville and Wilbur Wright are honored as the inventors of the first "flyable" plane. Nevertheless, their achievement was certainly dependent on already known cultural elements, and, as in the case of the automobile, the problem of making a successful "heavier-than-air flying machine" was being worked on over a broad spectrum of Western culture.

Such considerations only give a hint about the complexity of the problem regarding the relations between a group of people and their culture. Evidently the culture is "something" they have inherited from their forebears; from the latter they get their languages, religions, social customs, skills, tools, and, if sufficiently "civilized," their mathematics. But this "something" that has been passed on to them makes up their whole way of life; they not only have to live their lives in dependence on the culture they have inherited, but the only way they can make "progress" is to work from within the framework of that culture. Moreover, the changes or improvements that they can make are limited by the state of the culture as they inherit it. In every case, careful analysis shows that the culture was "ready" for the changes, which is particularly obvious in the case of the major changes. True, many minor changes, such as introduction of new words into the language, may be made by individuals—although even here the influence of the culture can usually be found (in the case of the term "normalcy," Harding would never have even thought of the word had it not been for the American economic and political situation at the time).

1.3 *Contrast between the "lives" of a culture and of a people.* Another important aspect of the link between a group of people and their culture resides in the fact that (barring catastrophic events) although the people involved are completely replaced by a new group—their descendants—in a certain finite time, their culture lives on, continuing its growth. In a sense, then, the culture is independent of the group possessing it, except insofar as the culture could not exist without the sustaining group. Anthropologists agree that the identity of the individuals forming the group is not an influential factor in the way in which the culture evolves. If one could imagine that in a given society each individual living at a given time had never been born, but had lost out to a potential brother or sister in the battle for life during the process of conception, the current state of the culture would still be substantially the same. Once we have arrived at this conclusion, the question naturally arises: are there "laws," analogous to those of the language elements of the culture, that "govern" the way in which the culture grows? Put another way, the question becomes: is it possible to formulate a theory of *evolution* of cultures analogous to the theory of biological evolution?

At this point, it is pertinent to state that in the modern era no culture exists alone; like the individual person, it is continually in contact with other cultures. And just as one person gets new ideas from another person through social contact, so does one culture absorb new elements from other cultures. If culture A makes a better mousetrap than culture B, and culture B finds out about it, then culture B will certainly copy the design—assuming, of course, that it has materials available, manpower to produce it, and so on, and that it is plagued by mice. Consequently, in seeking "laws" for cultural evolution, it would be necessary to take into account the influences to which a culture is subjected by other cultures.

2 PROCESSES OF CULTURE CHANGE AND GROWTH   As one would expect, the greatest changes take place in cultures that

(1) are in contact with the greatest number of other cultures and/ or which (2) have attained the greatest diversity of cultural elements. In case (1), *diffusion*[1] takes place, that is, concepts and customs pass from one culture to another; in case (2), the diversity gives rise to diffusion between cultural elements (as exemplified, for instance, by the business element of American culture exploiting the results of "pure" science or by the influence of television on the drama). In ancient times diffusion often occurred through the medium of trade or war; in modern times the instruments of diffusion are almost too numerous to enumerate; prominent among them are travel, printed media, radio, and television.

The likelihood of diffusion of a cultural element E from one culture to a neighboring culture, say from culture A to culture B, will naturally be greatly enhanced if greater efficiency in attaining the cultural objectives of B will ensue from adoption of the cultural element E. For example, in the case where A was the Spanish culture, E the horse, and B one of those Indian cultures that depended on hunting for its food supply, there was much greater efficiency to be attained by adopting the horse as a means for obtaining food (as well as for protection against enemies). But in the case of the Pueblo cultures, where a well-developed agriculture provided the food, use of the horse did not diffuse from A.

In other cases, the adoption of the new cultural element may be effected only partially, its advantages being attractive enough for certain parts of the host culture to bring about its diffusion thereto, but not for adoption by the total culture. Consider the case where culture B is our American culture and the cultural ele-

---

[1] Rather than become involved in a discussion of distinctions between such terms as "acculturation," "assimilation," and "diffusion," we shall generally use the last. Inasmuch as our discussion is oriented toward a particular cultural element—mathematics—and not intended as a treatise in anthropology, we are compelled to economize in terminology as much as possible, avoiding the refinements that would be needed in a general study of culture. [In Kroeber (1948, p. 426) we find: "When we follow the fortunes of a particular culture trait or complex or institution through its wanderings from culture to culture, we call it a study of diffusion."]

ment E is the metric system of measurement. Our scientific "sub-culture" has adopted the metric system, but the American culture as a whole has not. Insofar as the total American culture is con-cerned, this is an example of what anthropologists term cultural lag, since change to the metric sytem would ultimately bring dis-tinct advantages. This type of resistance to change, *cultural lag,* may be thought of as a type of conservatism, and it unquestionably has a survival value for a culture. In the words of Kroeber (1948, p. 257): "A culture that was so unstable and novelty-mongering that it could continually reverse its religion, government, social classification, property, food habits, manners and ethics, or could basically alter them within each lifetime . . . would scarcely seem attractive to live under . . . and it would presumably not survive very long in comparison or competition with more stable cultures, through lacking the necessary continuity."

A more overt obstacle to diffusion of a cultural element may be termed *cultural resistance.* This is observed especially in a case where adoption of the new element would replace an old element that serves the same purpose in the culture. Here there is cultural lag of the type described by Kroeber, but in addition there may be no evident increase of efficiency in attaining cultural objectives. An example of this is found in the area of religion— as any missionary could testify. Diffusion may, of course, occur through the medium of forceful imposition. Examples of this abound in the history of military conquest. And when the imposi-tion of the new element requires forcing out an old element serving the same purpose, the power of cultural resistance becomes par-ticularly striking. Thus one finds today, in the Pueblo cultures of the Southwestern United States, the ancient forms of the Pueblo religions coexisting and intermingled with the Roman Catholicism that was introduced by the Spanish conquerors more than 400 years ago!

An interesting phenomenon frequently observed in the case of diffusion accompanying military conquest is that in which the diffusion occurs in the reverse direction—from conquered to conqueror. This happened particularly in the case of mathematics. Although much unfortunate destruction accompanied the Moslem conquest of the seventh century, had not the conquerors assimilated so much of the mathematics of the conquered nations, much of the ancient Greek and Indian mathematical work, it can be conjectured, might have been forever lost.

It is evident, then, that if one seeks to formulate a theory of cultural evolution, such "forces" as diffusion, cultural lag, and cultural resistance will have to be taken into account. But to come back to the fundamental fact that a culture continues evolving even though its bearers, the people possessing the culture, die off, one generation after another, we may ask, what is it that makes this possible? For example, how was it possible for the airplane to grow from the humble type of "heavier-than-air flying machine" of the Wrights to the marvelous air carriers of today? Obviously because there was a medium by means of which the skills and knowledge of one group of scientists and technologists could be communicated to another, and this medium consisted not only of the written and spoken languages but of the symbols of mathematics, chemistry, engineering, and the like. All of the latter can be traced back to man's ability to symbolize (compare the remarks concerning "symbolic initiative" in Section 2 of the Introduction). Just how or why man developed the capacity for symbolizing, is not known. But it is this capacity that enabled man to *conceptualize* and to pass on his concepts to his colleagues and children. It is *symbolization* that not only makes cultures possible, but furnishes the means for their continuation and growth. (See White, 1949, Chapter 2.)

Anthropologists have constructed theories of evolution of culture whose scientific importance may prove to be as great as the more familiar theories of biological evolution (see, for example, Childe, 1951; Huxley, 1957, pp. 56-84; Sahlins and Service, 1960; White, 1959). Indeed, a complete history of man's evolution would today have to take into account his cultural evolution as well as his biological evolution, especially since it is evident that there have been influences of each on the other that could not be disregarded.

3 MATHEMATICS AS A CULTURE   In our discussion of the anthropological meaning of "culture" we pointed out the importance of the phrase "related by some associated factor," and we cited "common occupation" as one of the possible associative factors. One of the most important "common occupations" today is that of "doing mathematics." And those people who do mathematics—the "mathematicians"—are not only the possessors of the cultural element known as mathematics but, when taken as a group in their own right, so to speak, can be considered as the bearers of a *culture,* in this case mathematics.

However, it makes little difference whether one calls mathematics a culture or a cultural element. It is important, however, to observe that just as the culture of a people can be considered as forming an organic whole whose evolution invites study, so too can a culture or a cultural element such as mathematics. Moreover, it seems likely that a study of the evolution of a cultural element having such unique features as has mathematics may reveal forms and processes that are either not obvious, or do not have as great a significance, in the evolution of the cultures of whole societies. Thus, while it will be found that forces that are recognized as generally operative in the evolution of culture, such as diffusion, are quite as important in the case of mathematics, certain other forces, such as generalization, consolidation, and diversification (see Chapter 4, Section 4) have a special function in the evolution of mathematics. It can be expected, too, that symbolization will play a very important role in the evolution of mathematical concepts. For as a science becomes more abstract —and mathematics is notoriously abstract—its dependence on a well-devised symbolic apparatus becomes greater. Just how important such forces as diffusion, cultural lag, and cultural resistance prove to be in mathematical evolution can only be determined by taking into account the history of mathematics.

4 SYSTEMS OF NUMBER NOTATION   One of the most wonderful and ingenious of modern cultural elements is the decimal system of number notation. It is one of those things that we take for granted, much as we do the air we breathe. The average person probably knows less about its intrinsic properties than he does about the chemical composition of the air. Its evolution required the participation of many different cultures and a time span of more than forty centuries! And one of the most marvelous things about it is that with only ten digits $(0, 1, \ldots, 9)$ we can express *any* number whatsoever, no matter how small or large (an important consideration in these days of nuclear study and space travel).

Of what does it consist? All that it requires is a knowledge of two ingredients, namely *base* (which in the decimal system is the number 10) and *place value*. Consider the number 4325; in words, four thousand, three hundred and twenty-five. Why do we not say *three* thousand, *four* hundred, and twenty-five? This is because the *places* or *positions* in which the 4 and the 3 occur dictate the former—"dictate" in the sense that this is what we were taught in our youth. Perhaps the reader can recall his teacher saying that "In 4325, the 4 is in the thousands place, the 3 is in the hundreds place, the 2 is in the tens place, and the 5 is in the units place." And the meaning of the statement that "4 is in the thousands place" is that the 4 is really to stand for 4000; another way to state this is that the 4 in 4325 is to stand for $4 \times 10 \times 10 \times 10$, or, using the abbreviation that mathematicians use, 4×10^3. Similarly, the statement that in 4325 "3 is in the hundreds place" means that the 3 is to stand for 300—or $3 \times 10 \times 10$, which may be abbreviated to 3×10^2. The 2 in 4325 stands for 2×10— which for formal reasons we can also write as 2×10^1; and the 5 stands for just 5 units—which, as the reader who knows elementary algebra will recall, can be written 5×10^0 (since generally for any number n greater than 0, $n^0 = 1$).

Now, the number 43250 is not the same as 4325, because the 0 has moved each of the digits in 4325 one place to the left, so that now the 4 is in "tens of thousands place"—that is, the 4 now stands for $4 \times 10,000$ (the comma is inserted in 10,000 just for convenience in counting the number of digits and has no other significance). And $4 \times 10,000$ may be abbreviated to 4×10^4. Notice how important the symbol 0 is here; as will be seen later, the early Babylonians had not yet thought of such a symbol and were in the position we would be in if we had to denote both 43250 and 4325 by the symbol 4325, with the expectation that we would be able to infer which number was meant from the context in which the symbol 4325 was used.

Of course, today when we see the symbol 4325 we take it for granted that the number indicated is to be considered as in the base 10; that is, 4325 is $4 \times 10^3 + 3 \times 10^2 + 2 \times 10^1 + 5 \times 10^0$. But suppose that we were told that the symbol 4325 is intended to be *interpreted in the base* 7; what would this mean? It would mean precisely that we were to understand that 4325 means $4 \times 7^3 + 3 \times 7^2 + 2 \times 7^1 + 5 \times 7^0$. Since in our culture we have been brought up to use the decimal system, we would have to convert this expression into the decimal system before we could understand its meaning—that is, $4 \times 343 + 3 \times 49 + 2 \times 7 + 5 \times 1$—which is $1372 + 147 + 24 + 5 = 1548$. More generally, if we are to understand 4325 as denoting a number in the base b, where b is any number greater than 4, then 4325 would be $4 \times b^3 + 3 \times b^2 + 2 \times b^1 + 5 \times b^0$.

Now, why was the phrase "where b is any number greater than 4" inserted in the preceding sentence? This was because the number of basic symbols—the *digits*—needed is always *the same as the base number;* in the base 10 we have the ten digits 0, 1, . . . , 9. For the base 7 we would need only the digits from 0 to 6, and for the base 4, only the digits 0, 1, 2, 3 are needed. Of course this leads to the conclusion that the simplest case of all would be the base 2 (unless we revert to primitive tallying, which can be considered as equivalent to the use of the base 1). For the base 2 we need only two digits, 0 and 1. And using this base, the symbol 4325 has no meaning, since none of the digits 4, 3, 2, and 5 is one of the basic symbols 0, 1. However, we can write the *decimal* number 4325 in the base 2—or, as we usually say, in the *binary system* —and it will assume the extraordinary array 1,000,011,100,101 (again, the commas have no other significance except convenience in interpreting the symbol). To see this, let us start from the *right;* the first 1 is in units place and stands for $1 \times 2^0 = 1$. The next 1 is in the *third* place from the right and thus indicates the number $2^2 = 2 \times 2 = 4$ (remember that we are converting to *decimal* notation!). The next 1 from the right is in the *sixth* place from the right and so is $2^5 = 2 \times 2 \times 2 \times 2 \times 2 = 32$. Continuing in this way, the last 1 is in the thirteenth place from the right and consequently stands for 2^{12}, which is 4096 in *decimal* notation.

It should be obvious why we do not use the binary system for ordinary purposes; although this is not the precise way to phrase the statement intended, considering that *we have been indoctrinated in the decimal system* by our culture. What we should have said is that it should be obvious why we would not use the binary system for ordinary purposes even if we had any choice in the matter. For as numbers increase in size, considerably fewer digits are required to write them in the decimal system than in the binary. Consequently, if we had a choice between the two systems, we would certainly choose the decimal system in preference to the binary.

As a matter of fact, if we had had a choice in the matter, we probably would not be using either the decimal or the binary system—for ordinary purposes, that is. One can make quite a good argument for the *duodecimal* system, which has the base 12. Of course, this system requires twelve digital symbols: the ten digits

0, 1, . . . , 9 already in use for the decimal system can be used, together with two additional symbols, X (called dek) and ε (called el). For most purposes it is quite superior to our decimal system and has found enough uses so that tables of logarithms and other functions have been supplied for it.[2] It is not unlikely that if nature had supplied us with six digital members on each hand, instead of five, we would be using the duodecimal system today for most purposes.[3]

For scientific purposes the binary system is constantly used. Not only does it lie at the base of computer operations, but is most useful in mathematical theory.

On the presumption that the reader who has found anything informative in these remarks on number notation will also wish to know about its extension to representation of fractional numbers, an exposition of the meanings of numbers such as 32.4126 is added here. The decimal point is used to separate the integral part of the number—32—from the so-called decimal part—4126. As already described, the integral part stands for $3 \times 10^1 + 2 \times 10^0$. To interpret the decimal part, 4126, one proceeds just as in reading a thermometer, that is, by using minus signs: $4 \times 10^{-1} + 1 \times 10^{-2} + 2 \times 10^{-3} + 6 \times 10^{-4}$. The reader who had elementary algebra may recall that a symbol such as a^{-b} means the same as $1/a^b$ (except when $a = 0$). So 10^{-1} is $1/10$, 10^{-2} is $1/10^2 = 1/100$, and so on. For other bases, the same procedure is used. Thus, in the base seven, the symbol 32.4126 stands for $3 \times 7^1 + 2 \times 7^0 + 4 \times 7^{-1} + 1 \times 7^{-2} + 2 \times 7^{-3} + 6 \times 7^{-4}$, which becomes $3 \times 7 + 2 + 4/7 + 1/49 + 2/343 + 6/2401$ in decimal notation.

Now, in order that the reader not suspect "cheating," a comment should be added concerning our assertion that the decimal

[2] There exists an organization for the propagation of the duodecimal system called The Duodecimal Society of America.

[3] Propaganda for the base 8—resulting in the "octonary system"—can be found in the literature. See Tingley, 1934, for instance. This base would be found to be quite convenient in working with stock market quotations!

system (the same holds also for other systems—such as duo-decimal and octonary) is capable of expressing *any* number, no matter how large or small. From what was said, it should be clear that it will express "counting numbers"—that is, 1, 2, 3, and so on—just as large as wished. Also, using the decimal point and the notation explained in the previous paragraph, one can obtain just as small numbers as wished—for instance, 0.1, 0.01, 0.001, 0.0001, and so on (using more and more zeros). But how can one express the common fraction $\frac{1}{3}$ in decimal notation? Or "worse" still, how about $\sqrt{2}$? These numbers lead to so-called infinite deci-mals, and for a proper understanding of these the exposition will be delayed until Chapter 3 (in which the "real numbers" are dis-cussed). All that is needed until then is an understanding of base and place value (sometimes called positional notation).

CHAPTER ONE

Early Evolution of Number

1   INCEPTION OF COUNTING   A well-known school of mathematical thought[1] holds that all mathematics should be founded on the counting numbers—1, 2, 3, and so on—technically known as the *natural numbers.* Certainly from an evolutionary standpoint this view is justified, for all the records—anthropological and historical—show that counting and, ultimately, numeral systems as a device for counting form the inception of the mathematical element in all cultures that have been unaffected by diffusion. Anthropologists have found some form of counting in all primitive cultures, even among the most primitive that have been observed, even though it may have been represented only by a few number words.

1.1   *Environmental stress, physical and cultural*   Evidently the rudiments of counting form a *cultural necessity,* hence what anthropologists call a cultural universal.[2] In addition to the inevitable physical environment and the problems that it presents, man's social and cultural environments also demand recognition of the difference between "oneness" and "twoness." Even animals, which have no number system, seem to have this ability.[3] In man,

---

[1] Intuitionism; see Chapter 5, Section 2.2.

[2] The anthropologist George P. Murdock listed "numerals" as one of 72 items that occurred, so far as was known, in every culture known to history or ethnography.

[3] See Conant's classical work (1896, pp. 3-6) on number for a quotation

the culture-building animal who is equipped with a symboling faculty, the environmental stress resulting from cultural pressures —what I shall call *cultural stress*—induces a more mature facility, which, using numeral symbols, evolves into "threeness," "fourness," and so on beyond the capacities of nonsymbol-using creatures. This factor of "environmental necessity" has operated throughout the evolution of mathematics, and all the evidence shows that it played a major role in the inception of those cultural elements that later, in some cultures, became a recognizable mathematics.

True counting is a process whereby a correspondence is set up between the objects of the collection to be counted and certain symbols, verbal or written. As practiced today, the symbols used are the natural number symbols 1, 2, 3, and so on. But any other symbols will do: tally marks on a stick, knots on a string, or marks on paper such as

$$\text{||||| \quad ||}$$

are all sufficient for elementary counting purposes. Counting is, then, a *symbolic process* employed only by man, the sole symbol-creating animal.

Perception of inequality in the sizes of two collections is not, of course, necessarily dependent on counting. This is particularly the case where one of the collections is so much vaster than the other that the eye can perceive this fact at a glance. Only when the stresses evoked by the growing complexity of a culture become great enough to induce it does the more refined method of *counting* emerge. One is reminded here of the situation in some primitive tribes relative to color words. That certain tribes have only one

and discussion of remarks of Sir John Lubbock regarding number sense in animals and insects. We shall not enter into a discussion of the nature of this ability, except to remark that in the absence of a symbolic faculty, it can hardly be called counting and is necessarily limited to collections having only a few elements; and that, if evidence to be cited later regarding precedence of a number concept by ideographs is conclusive, it cannot be considered as a use of number per se.

term for green and blue was originally considered to mean that they were unable to distinguish between the two colors; now it is recognized that they actually could distinguish between them visually, but that the cultural need for such a distinction was not strong enough to compel a verbal equivalent for the differentiation.

That the necessity for the emergence of a counting facility in every culture is rather generally admitted, at least tacitly, has recently been made manifest in the numerous popular and semi-popular articles that have been written regarding establishment of radio contacts with possible culture-building animals on planets other than our own. Assuming the existence of cultures elsewhere than on our earth, is there any communicable cultural element that *we* are acquainted with and that we may confidently expect to find simulated in such cultures? The opinion seems to be unanimous that there is, and that the most likely one is the existence of counting processes, hence also the existence of the concept of natural number and of modes of representing it symbolically.

For example, an article in *Science* (December 25, 1959) relating to the then-proposed attempts by the National Radio Astronomy Observatory at Greenbank, West Virginia, to make interplanetary radio contacts, comments: "What kind of signals might we expect [to receive]? Radio astronomers agree that pulses to communicate prime numbers or some simple arithmetical problems might be suitable."[4] Or consider the following excerpts from a talk by the well-known popularizer of elementary mathematics, Lancelot Hogben, before the British Interplanetary Society (quoted from *Time,* April 14, 1952):

". . . What can earthlings say that their Extra-terrestrial

---

[4] An amusing, though perhaps irrelevant, side remark accompanies this: "If you ask radio astronomers why we don't start to broadcast, you learn that they think fiscal authorities would not approve. This leads to an unhappy thought: May not other civilizations have evolved analogous fiscal authorities? And may they not likewise be waiting in silence before they give response?" See G. DuShane, "Next Question," *Science,* Vol. 130 (December 25, 1959) p. 1733.

Neighbors will understand? Let's begin, said Hogben, by some small talk about numbers, *whose properties do not vary from planet to planet* [Italics ours]. . . . Probably the Neighbors passed through a similar stage [of the development of counting] and have records of it. So Hogben's first message into space would be an equation in simplified numerals:

'I plus II plus III equals IIIIII'

"The numbers are 'dashes' (single strokes repeated), and the plus signs and equal sign are 'flashes.' [By "flashes" Hogben means easily recognized groups of radio signals, rather like the letters of the Morse code.]

"When the Neighbors have heard this equation, repeated often enough, they ought to understand its meaning. By taking it apart, they can learn the first words of the interplanetary language. More complicated equations will teach them more words."[5]

Whether counting started in a single, prehistoric culture and spread thereafter by diffusion[6] or developed independently in various cultures (as seems most likely), is perhaps not too important for our purposes, interesting as it may be to speculate thereon. The scarcity of knowledge about modern man's forebears has not greatly impeded the study of biological evolution; and since it seems impossible to find out *when* man developed counting, relative to his biological origin and geographical spread, we may as well get on with what we know from the archaeological and historical records. Even use of the word "started" seems inadmissible indeed, inasmuch as counting could hardly have "started" in either the individual or the historical sense. Even if it evolved in a single primitive center, it did *evolve* and, as in the case of many cultural elements, the event would allow *dating* only by a convention (the like will be found to be true of mathematical concepts in general; see the discussion of the evolution of calculus in Chapter 4, Sec-

---

[5] Courtesy *Time;* © Time, Inc., 1952.
[6] As maintained by A. Seidenberg (1960).

tion 3, which convention "dates" as commencing with Leibniz and Newton, but which actually evolved from prior "beginnings"). Number words are usually found among the earliest word forms in a written language. "Both in Sumer and in Egypt there are documents, using a conventional system of numeration, that are earlier than the earliest extant examples of writing" (Childe, 1948, pp. 195-196).

1.2 *Primitive counting*   A great deal has been written about early numerals. Anthropologists have accumulated an abundance of details concerning *numeral systems* from various cultures, and a thorough descriptive study and analysis of them would require volumes.[7] So far as present purposes are concerned, however, it will suffice to point out their most significant characteristics.

1.2a *Distinction between "numeral" and "number"* First, it should be made clear how the terms "number" and "numeral" will be employed. By "numeral" will always be meant a *symbol;* this seems to be in accord with general usage. The word "number" will usually indicate the *concept* symbolized by a numeral or numerals. "Number" has come to have both an individual and a collective implication; its use in the phrases "the number 2" and "the nature of number" illustrates this. There is nothing peculiar about this; the same occurs in the use of the term "man" —for instance, "that man is an American" and "man is an animal" —and of many other nouns. Usually the context will make clear which usage is intended—ordinarily the term will be used in the individual sense, however, Much more important—and troublesome—is the actual characterization of the concept. The question "What is a number?" has led to countless discussion and argument.

1.2b *Distinction between "cardinal" and "ordinal"*   In

---

[7] See the works by V. G. Childe, A. L. Kroeber, and E. B. Tylor cited in the Bibliography. See also the classic work of Conant (1896), as well as works of S. Gandz, K. Menninger, O. Neugebauer, G. Sarton, Thureau-Dangin, A. Seidenberg, and D. Smeltzer.

everyday usage, numbers have both a "cardinal" and an "ordinal" sense. A cardinal number answers the question "How many?"— as in "2 dollars" or "2 days." An ordinal number indicates not only how many, but also answers the question "In what order?" or "In what position of the given order?" For example, the day of the month or the number of a theater seat is really an ordinal— we may write "January 3" but we mean "the third of January."

It is interesting to note that when we count a collection of things, we actually mix the two concepts; we may only be interested in *how many*—the cardinal number—but we count ordinally—"1, 2, 3, 4, ..." That is, we order the objects in the process of ascertaining the cardinal number of the collection.

1.2c *"Two-counting"*   Many scholars point to the words used to indicate ordinals as evidence for the assertion that the earliest counting was "2-counting"; that is, that the earliest numerals were of the sort "one," "two," either accompanied by no further number words and employing "many" for more than two or using "two-one" for three, "two-two" for four, and so on. It is striking to observe the prevalence, in modern languages, of different forms for the two words "first" and "second" and the succeeding words "third," "fourth," "fifth," and so on. The English forms that we have just used are illustrative of this—we say "first," not "oneth," and "second," not "twoth."[8] Additional linguistic evidence consists of the dual forms of the plural in ancient languages; instead of just singular and plural forms of nouns one finds what may be called "singular," "dual," and "plural"—the dual form of the noun indicating two of the objects denoted and the plural form indicating three or more. As one author has remarked,[9] "It is as if we were to write cat, catwo and cats when referring to one, two and more than two cats." Tylor remarks (1958, p. 265) that "Egyptian, Arabic, Hebrew, Sanskrit, Greek, Gothic, are examples

---

[8] For examples in other languages see Tylor (1958, pp. 257-258) and Smeltzer (1953, pp. 5-8).

[9] Bodmer, as cited in Smeltzer (1953, p. 6).

of languages using singular, dual and plural numbers; but the tendency of higher intellectual cultures has been to discard the plan as inconvenient and unprofitable, and only to distinguish singular and plural. No doubt the dual held its place by inheritance from an early period of culture, and Dr. D. Wilson seems justified in his opinion that it 'preserves to us the memorial of that stage of thought when all beyond two was an idea of indefinite number!' "[10] [Quoted with publisher's permission.]

As remarked above, the necessity for adapting to both cultural and physical environmental stresses must have compelled the recognition of "oneness" and "twoness"—and, ultimately, of the "beyond" or "many." Conant remarks (1896, p. 76): "It is to be noticed that the Indo-European words for 3—*three, trois, drei, tres, tri,* etc., have the same root as the Latin *trans,* beyond." One can cite worldwide instances of primitive cultures in which the numeral words seem to be confined to the equivalent of "one, two, many" where "many" indicates "three or more."

1.2d *Tallying; one-to-one correspondence*   It is interesting to speculate on the origin of higher numerals. For example, *tallying* formed an early type of enumeration. Evidence exists of the use of tally sticks, in which notches were cut, even in Paleolithic times. Knots tied in a string were another common form of tallying. And the custom of making marks on some convenient substance—such as sand, a cave wall, clay that was subsequently baked or sundried, or papyrus—was widespread. The most advanced form of tallying occurs in the many forms of the abacus, used both in ancient and modern cultures. One can surmise, also, that the introduction of tallying by writing ultimately led to ideographs, that is, numeral symbols.

It is important to notice the psychological element that enters into tallying, that is, the intuition of one-to-one correspondence.

[10] Tylor refers here to D. Wilson, "Prehistoric Man," p. 616. It is interesting to note that in other grammatical categories the triad of forms is still used; thus "good, better, best."

When sheep are counted by passing them one by one through a gate, while simultaneously a pebble is placed in a heap for each sheep that passes, the counter is intuitively comprehending what the modern mathematician calls a (1-1)-correspondence.

The choice of primitive number words is another indication of such intuitive comprehension of (1-1)-correspondence. Those who agree that finger-counting preceded the use of number words cite how universal has been the use of the word "hand" for "five." Similarly 10 was called "two hands" or (in cultures that extended finger-counting to toe-counting) "half a man," whereas 20 might be called "one man," for instance. Counting with fingers (and toes) clearly involves intuitive recognition of (1-1)-correspondence. Likewise, the use of "eye" for 2 reveals an understanding of the (1-1)-correspondence between one's collection of eyes, say, and a pair of twins. Modern mathematics bases the notion of cardinal number on this idea of (1-1)-correspondence. But the journey from the early intuition to the modern concept is a long and meandering one, during which many dead-end bypaths were followed.

1.2e *Number categories; adjectival forms*   One such bypath may have been the use of different number words for differing categories of objects. Although this may be considered as an intermediate stage in the general evolution of number, it was probably not a stage passed through in every culture (possibly because of diffusion). That its occurrence was widespread is shown conclusively by the evidence. Frequently cited are the findings among the British Columbia tribes[11] by the late anthropologist Franz Boas. A modern form of this phenomenon persists in the Japanese language, where different word forms are used for numerals in the range 1 to 10 relating to persons, dishes, and pencils, for instance. These forms derive from two sources—the ancient Japanese and the Chinese—from the latter by diffusion. For ordi-

[11] See the tabulations on pp. 87-88 of Conant (1896).

nary counting, the suffix "tsu" is attached to the ancient Japanese root; thus "itsu"—5—becomes "itsutsu." But for counting long, slender objects (pencils, poles, trees) a suffix varying between "hon," "bon," and "pon" is used. And here, curiously, it is at-

No.	Count-ing.	Flat Objects.	Round Objects.	Men.	Long Objects.	Canoes.	Measures.
1	gyak	gak	g'erel	k'al	k'awutskan	k'amaet	k'al
2	t'epqat	t'epqat	goupel	t'epqadal	gaopskan	g'alpēeltk	gulbel
3	guant	guant	gutle	gulal	galtskan	galtskantk	guleont
4	tqalpq	tqalpq	tqalpq	tqalpqdal	tqaapskan	tqalpqsk	tqalpqalont
5	kctōnc	kctōnc	kctōnc	kcenecal	k'etoentskan	kctōonsk	kctonsilont
6	k'alt	k'alt	k'alt	k'aldal	k'aoltskan	k'altk	k'aldelont
7	t'epqalt	t'epqalt	t'epqalt	t'epqaldal	t'epqaltskan	t'epqaltk	t'epqaldelont
8	guandalt	yuktalt	yuktalt	yuktleadal	ek'tlaedskan	yuktaltk	yuktaldelont
9	kctemac	kctemac	kctemac	kctemacal	kctemaetskan	kctemack	kctemasilont
10	gy'ap	gy'ap	kpēel	kpal	kpēetskan	gy'apsk	kpeont

Tsimshian number characters. (L. L. Conant, *The Number Concept*, Macmillan, New York, 1896.)

tached to one or the other Japanese or Chinese roots but not exclusively to either; thus for 5 "go-hon" is used, where "go" is the Chinese-derived root, but 7 becomes "nana-hon," where "nana" is the ancient-Japanese-derived root.

Such word forms also point to a descriptive—that is, adjectival—use of number words. And it seems not unlikely that number words passed through such a stage in most of those cultures that were not affected by contacts with other cultures. The uses in modern cultures of number words are both adjectival and objective; in "two trees" the "two" is an adjective, whereas in "the number two" it is a noun. Dictionaries recognize both uses. But although every person in the modern English-speaking cultures knows what he means by "two trees," it is doubtful if very many know precisely what they mean by "two" in the expression "the number two." Probably the average person would resort to writing "2" and pointing! This prompts one to speculate that number words may never have been used as nouns until some ideograph, such as "2" or (more likely) "||," had been in use for some time. Existence of an ideograph seems eventually to lead to objective status.

Many marvelous and ingenious word forms were devised in primitive cultures under the pressures of cultural stress. These can be found in profusion in works already cited. Having installed words for "one" and "two," a culture may later proceed to call 3 "two-one," 4 "two-two," and so on, thereby possibly setting forth on the road to the ultimate use of a binary system. Rudiments of addition, subtraction, and even multiplication have been found in primitive word forms; for example, 9 has been found to be represented by a word indicating "10 minus 1," 8 by "10 minus 2," and even 6 by "10 minus 4" (Ainu) (see Conant, 1896, pp. 44 ff). (An analogous phenomenon later occurred in more advanced cultures in which, even though numeral ideographs were present, words had to be introduced to compensate for a lack of *algebraic* symbols).

Since the most widely prevailing bases are quinary, decimal, or vigesimal, it is likely that the use of finger-counting dictated the ultimate choice of base. As might be expected, there are abundant examples in which a culture's number words were affected by diffusion—as, for example, in the case of the Japanese.

2 WRITTEN NUMERAL SYSTEMS   So long as the symbols for numbers were only verbal, no great advance in the evolution of number seems to have been made. This is not to say that words for counting large collections were not introduced or were incapable of being introduced, since in some cultures they were; but the ultimate advance in the conceptual status of number was promoted by the introduction of ideographs. This is not surprising; simple arithmetic can hardly be developed very far without ideographic symbols.

2.1 *Sumerian-Babylonian and Mayan numerals; place value; zero symbol* In what may be considered the mainstream of our own cultural background, the Mesopotamian civilization, a fortuitous event in the evolution of arithmetic symbols seems to have occurred through the adoption of the Sumerian script by the conquering Akkadians. (The Sumerian *language* was also adopted, but as an official or sacred language, much as Latin was adopted in medieval days by our own forebears.) This is an example of the type of diffusion in which a culture that absorbs or is superimposed on another takes over cultural elements from the latter (*cf.* Preliminary Notions, Section 2). A by-product of this was the *symbolization* that resulted—possibly one of the most momentous events in the evolution of early mathematics. So long as mathematics is couched in the language of ordinary discourse, it can hardly advance very far. Modern ideograms such as $+$ and $=$ furnish a good example.

It was a fortunate historical accident that led to the Babylonian introduction of ideograms. As Waismann remarks (1951, p. 51), "mathematical symbols do not lie . . . in the direction of the natural development of language"; and "out of the letter writing which followed the phonetic pictures of the words, how could the concept symbols of mathematics spring up? . . . In Egypt, with its historical continuity, this step did not take place. In Babylon, where two entirely different cultures, the Sumerian and the Akkadian (with languages of basically distinct grammatical types) are superimposed, the path for such a formal development was cleared. . . . [T]hrough the contact of these distinct languages there arose the possibility of writing a word either by syllables or by ideograms. In the Akkadian texts both modes of writing are arbitrarily used in turn; thereby arose the possibility of writing the mathematical concepts (quantities and operations) ideographically and of attaining a language of formulae, while the remaining text is written in syllables." [Quoted with publisher's permission.]

And Childe states (1948, p. 204), "The Babylonion texts . . . from 2000 B.C. employ a very explicit terminology. Indeed, the Babylonians were well on the way to creating a mathematical symbolism that would materially accelerate calculation. First, the technical terms for several operations were words of one syllable expressed by a single cuneiform sign. Then the Babylonians, though they spoke a Semitic tongue, used the old Sumerian terms for operations like 'multiplied by,' 'find the reciprocal of.' Finally, many of the technical words were written as ideograms instead of being spelt out. . . . The later the texts, and therefore the deader the Sumerian language, the more Sumerian terms and ideograms were used. They became quite abstract symbols freed from the concrete notions of 'nodding the head' or 'breaking away' that inhere in Egyptian times." [Quoted with publisher's permission.]

Awareness of the achievements of Babylonian mathematics, incidentally, has been acquired only comparatively recently. Hitherto the greater emphasis had been placed on Egyptian mathematics. Thanks mainly to the work of historians such as O. Neugebauer and F. Thureau-Dangin, it is now known that, except for possibly certain geometric rules for mensuration, the Babylonian achievements in general mathematics far outstripped the Egyptian (indeed, many of the latter were evidently the result of diffusion from Babylon). The reasons for the Babylonian advances were manifold, but probably lie chiefly in the fact that Mesopotamia was geographically so situated as to encourage both conquest and trade (hence accelerated diffusion), while ancient Egypt was somewhat culturally isolated.

2.1a *The bases* 10 *and* 60   Frequently the adoption of a new form of a cultural element will not result in the elimination of the old form, both cultural lag and possibly cultural resistance occurring (*cf*. Preliminary Notions, Section 2); or, alternatively, *both* forms may persist if greater utility results. Today, for example, we still find uses for the Roman numerals. It is not surprising, therefore, to find in early Babylonian records a mixture of bases —10 and 60. The base 60 was commonly used by the Sumerians, and the base 10 was probably used by the Akkadians. After the Akkadians became dominant, evidently the use of the already existing Sumerian numerals resulted in preservation of the base 60, although the base 10 persisted in ordinary discourse. "Prior to the end of the third millennium (B.C.), the sexagesimal system has almost completely disappeared from the common usage" (Thureau-Dangin, 1939, p. 108). However, the sexagesimal system continued to prevail in scholarly work, particularly astronomy, probably because of its extension to the expression of fractions (see below).

Neugebauer (1957, pp. 17 ff) states that a clay tablet containing hundreds of astronomical numbers, all written in sexagesimal numerals, might end with a "colophon" giving the name of the scribe and date of writing, with the latter in the base 10 (more precisely, in powers of 10 rather than in powers of 60). He points out that "it is only in strictly mathematical or astronomical contexts that the sexagesimal system is consistently applied. In all other matters (dates, measures of weight, areas, etc.) use was made of mixed systems which have their exact parallel in the chaos of 60-division, 24-division, 10-division, 2-division which characterizes the units of our own civilization. . . . [M]any modifications of number symbols were in use for different classes of objects, such as capacity measures, weights, areas, etc. Among these a clear decimal system has been recognized with signs for 1, 10, and 100. . . . Another system proceeds sexagesimally, at least partially. . . . Variations of these systems, both decimal and more or less sexagesimal, can be established at different localities."[12]

---

[12] From "The Exact Sciences in Antiquity," by O. Neugebauer, published by Brown University Press, 1957. This, and later quotations from the same work, are used by permission of the publisher.

Regarding the *origin* of the unusual base 60 in Sumer, many conjectures have been made. The influence of Chinese contacts (the base 60 also occurred in China) has been suggested. F. Thureau-Dangin (1939, p. 104) observes that "apparently, the unit of 60 has been incorporated in a system of numeration which was still in the process of formation, which had already the unit of 10, but had not yet, had never had the unit of 100." And, "Potentially, there is already contained in the series 1, 10, 60, the whole system of Sumerian numeration, which is, properly speaking, not a sexagesimal system, but a system built up on the two alternating bases of 10 and 60."

According to Neugebauer, "In economic texts units of weight, measuring silver, were of primary importance. These units seem to have been arranged from early times in a ratio 60 to 1 for the main units 'mana' (the Greek 'mina') and shekel. Though the details of this process cannot be described accurately, it is not surprising to see this same ratio applied to other units and then to numbers in general. In other words, any sixtieth could have been called a shekel because of the familiar meaning of this concept in all financial transactions. Thus the 'sexagesimal' order eventually became the main numerical system . . ." (1957, p. 19).

Others have conjectured that the base 60 arose from astronomical considerations; still others that it was deliberately selected because of its having so many convenient factors (2, 3, 4, . . .). A rather complete catalog of these various conjectures may be found in Thureau-Dangin (1939, pp. 95-108).

The survival of the sexagesimal system in Babylonian science was undoubtedly due to its extension to fractions. Our modern treatment of fractions takes two forms: the *rational* fraction (i.e., the quotient of two integers) and the *decimal* fraction; in the former, one quarter is written as ¼ and in the latter as 0.25. Use of the decimal form has the advantage of unifying operations with integers and fractions. Thus multiplying by 0.25 is, except for the placing of the decimal point, equivalent to multiplying by 25, and dividing by 125 is equivalent to multiplying by 0.008. Not only is this important for machine computation, where division can be replaced by multiplication, but it permits treating numbers and operations in a unified fashion in all theoretical considerations. However, most ancient civilizations, and this includes the Sumerian as well as the Egyptian and Greek (except in the case of Babylonian and Greek astronomical work) used only the rational forms insofar as they had any fractions.

Example of a Babylonian clay tablet. (The University Museum, University of Pennsylvania.)

2.1b *Place value in the Babylonian and Mayan numeral systems* Now, the decimal form of fractions is based on place value notation, in which the value of a digit depends upon its position relative to the decimal point (in 25, the 2 stands for 20, that is, 2 times 10, and in 0.25 it stands for two tenths or 2 times 10^{-1}). (See Preliminary Notions, Section 4). Neugebauer counts the invention of place value notation as "undoubtedly one of the most fertile inventions of humanity" that can "be properly compared with the invention of the alphabet as contrasted to the use of picture-signs intended to convey a direct representation of the concept in question" (1957, p. 5). There has naturally been much speculation regarding its origin (it occurred in both the Babylonian and Mayan cultures).

The Babylonians wrote with a reed stylus on soft clay tablets, which were afterwards baked or dried in the sun.[13] For numerals, the Sumerians used reeds with circular ends in two sizes. The symbol for unity was made by pressing the smaller end from a slanted position, producing a sort of half-moon, and the symbol for ten was obtained by pressing from the vertical, resulting in a full-moon form. Using the larger end, the half-moon symbol symbolized 60 and the full moon 3600. Symbols for the integers from 2 to 9 followed the primitive tally system of repetitions of the symbol for unity [although according to Thureau-Dangin (1939, p. 106), "The scribes made abundant use of the subtractive method expressing 9 as 10 minus one"]. On the other hand, according to Neugebauer (1957, p. 19), 100 in the commonly employed *decimal* system was symbolized by the larger full moon. Common to all the variations of decimal and sexagesimal systems in various localities were "the existence of a decimal substratum and the use of bigger symbols to represent higher units. *This latter fact is obviously the root for the development of the place value notation* [Italics mine: RLW]. When the script slowly became simplified and standardized, the distinction between bigger and small signs of the same type disappeared. Whereas originally one big unit, meaning 60, and one 10 symbol were written to denote 60 + 10, later a single '1' followed by a 10 was read 70, in contrast to a 10 followed by 1 meaning 11."

In another connection, Neugebauer (1960) writes: "In the highly developed economic life of this period, a notation for the recording of monetary transactions originated in which larger and smaller units of weights of silver were expressed by simple juxtaposition of numbers, denoting units of different values, similar to the notation $5.20 as distinguished from $20.50. Thus the arrange-

---

[13] For a very engaging discussion of this process, especially of why and how it was employed, see Chiera, 1938, Chapters I, II.

ment of the numbers determined their relative value; in our example the ratio of dollars to cents is 1 to 100, in the Babylonian monetary system it happened to be 1 to 60. This notation was then extended to numbers in general, so that one ended up with a 'sexagesimal place-value system.' "[14]

Interestingly enough the anthropologist A. L. Kroeber had arrived independently at a similar theory. He pointed out (1948, pp. 470-471) that the Mayan "calendar system operated somewhat like the Mesopotamian weight measures in providing a scheme of multiplied ranks or orders, which must have gone far to suggest position values for numbers. Thus, much as in Mesopotamia, [where] 180 grams made a shekel, 60 shekels a mina, and 60 minas a talent, so with the Maya 20 days made a 'month,' 18 months a 'year,' 20 years a 'lustrum' or katun, 20 katuns a cycle. This regularity inevitably engendered habits of designating by position instead of by name, especially when there was abundant numbering or calculating to be done [cultural stress] . . . much as in British bookkeeping with £s.d. The situations were really quite parallel when a Neo-Babylonian wanted to designate a weight of 2 talents 6 shekels; a Maya, two 'years' and six days; or a Londoner, 2 pounds sixpence. And from these situations it is only a step, so far as manner of operation is concerned, to our writing the abstract value 206 or a Babylonian writing his '206' to denote 7206. While we do not know too much about how the Maya carried out their calculatory operations, we can assume as likely that when they wanted to add several time intervals, or when they substracted one date from another to learn the length of the elapsed period, they put the days, months, years, and so on in columns, one under the other, much, say, as when we want to subtract 206 from 773. Therewith we have what we may call actual 'positioned operation'; and if there is very much of this going on, it seems likely to tend to force the hand of the operator toward devising some means [a zero symbol] of designating absence, especially of internal units—of minas or months or shillings or sixties or tens, as the case may be." [Quoted with publisher's permission.]

Other cultures came close to place value systems without actually achieving them.

One may be tempted to wonder why, if representatives of numbers in the two bases 10 and 60 coexisted in Babylonian number systems, the place value concept used in the sexagesimal system was not extended to numerals in the base 10. Many conjectures can be made regarding the reasons for this. Certainly the cultural stress that was involved in the evolution of place value in the sexagesimal system could not have been equally operative in the case of the base 10. The importance of place value notation lies in its capability for expressing numbers as large as one wishes, or as small as one wishes, in terms of the same basic digits. This was important in Babylonian astronomy for the construction of tables, but in other areas, such as the marketplace, there was no comparable need.[15] Moreover, the variety and complexity of numeral systems, apparently used for various purposes (see Neugebauer, 1957, p. 17, for instance), may have brought a kind of cultural lag or cultural resistance into play. One can even conjecture that it did occur to a temple scribe now and then that what we call place value could be applied to other systems. However, it seems doubtful whether there was current, at that time, any such thing as a *concept* of place value; it was more likely just a *device* that the Babylonian mathematicians had found useful. As we shall see later, the possibility of utilizing place value for expressing numbers in the base 10 was apparently realized by various individuals many years before the work of Stevin in the sixteenth century; to

---

[15] We shall observe an analogous phenomenon in the Greek culture, where the numerals in common use made no use of the place value notation, but the astronomical tables continued to use it. (See Chapter 2, Section 3.)

these individuals, however, it was probably still only a "device," and Stevin is perhaps properly credited with bringing out the concept proper.

It is probable that the invention of place value systems represents a natural kind of progression in the evolution of numeral systems, brought about by cultural stress in the form of a need for symbolizing arbitrarily small or large numbers; or they may be a natural extension of such systems of measurement as the Babylonian monetary or Mayan calendar systems. Nevertheless, not all cultures would possess numeral systems that pass on to this stage of progress, any more than in biological evolution did all life-forms pass on to the amphibian and mammalian stages; most probably would not, as witness the later (and higher) cultures of China, Greece, and Rome, for instance.

2.1c *Zero symbols* Further light may be shed on this by a consideration of the zero symbol, already suggested in the quotation from Kroeber above. On the Old Babylonian tablets (c. 1800 B.C.), although place value is already in use, no zero symbol seems to have been used; instead, an empty space might be left between numerals. However, since an empty space at the end of an integer is not noticeable, it evidently became customary to let the context indicate the meaning. Thus the 1 symbol might stand for 1 or 60. It might be expected, in this situation, that cultural stress would promote invention of a zero symbol and, indeed, would ultimately make it inevitable. And so it seems to have been, since it occurred in both the Babylonian (Seleucid period) and Mayan numeral systems.

The Babylonian zero was apparently only a symbol indicating absence of a numeral and was used only medially, not at the right of the number; thus 402 and 42 would be distinguishable, but not 420 and 42 (whose meanings would have to be determined by the context). (See Neugebauer, 1957.) The Mayan zero symbol resembled a closed fist—suggesting that it evolved from a finger-counting stage. Sanchez (1961) gives an interesting analysis

of what the Mayan arithmetic may have been like, with examples
of addition, multiplication, and the like all worked out. As in the
case of Tannery's experience with Greek arithmetic (see Section
2.2a of this chapter and Section 3.2 of Chapter 2), Sanchez also
concludes that doing arithmetic with Mayan numerals would have
been quite simple (provided, of course, that the Mayans carried
out their operations in the presumed manner).

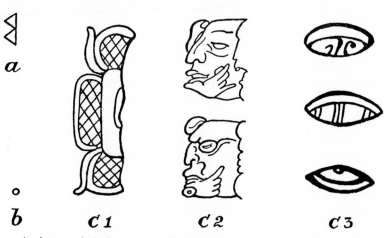

Ancient symbols for zero—a, Babylonian cuneiform; b, India; c, Maya;
c1, monumental inscription style, c2, face inscriptions, c3, written in books.
(A. L. Kroeber, *Anthropology*, Harcourt, Brace, and World, Inc., New
York, 1948.)

The possibility of diffusion operating between the Babylonian
and Mayan cultures seems nil, since the two inventions were apparently almost simultaneous (the Babylonian c. 400-300 B.C.
and the Mayan about the beginning of the Christian era—although
possibly much earlier) and occurred in areas so widely separated
as to render the passage of such cultural elements from one culture to the other possible only over a longer time interval than is
indicated by the records. It is interesting to note that in the same
connection Kroeber (1948) remarks: "Corresponding somewhat
to the Babylonian addiction to operations with a multiplicative
scale of monetary weight units, and of the Mayan with time units,

as the antecedents of their respective inventions of zero, is the Hindu use of the ancestral 'Arabic' digits from 1 to 9 for several centuries before they added a dot as a sign to denote that a certain position was vacant." The latter was at a later date (c. 500 A.D.) and in a culture not so far removed from Mesopotamia as is Latin America, so that one cannot be so certain of the absence of diffusion (many scholars seem to favor the introduction of a zero symbol in India as independent, for example, Kroeber, 1948; but see Section 2.3). Kroeber asserts (1948) that the evolution of zero "has evidently always been made against considerable psychological resistance [cultural resistance?]. The 'natural' or spontaneous thing [after inventing natural number symbols] is not to go on and add one also for nullity but to let nothing denote nothing." [Quoted with publisher's permission.]

2.1d *Sexagesimal fractions* Most remarkable, perhaps, was the extension of the Babylonian numeral notation to sexagesimal fractions (*cf.* Section 2.1a). Moreover, this facility existed in the Old Babylonian (before 1800 B.C.) period (although there was no analogue of a decimal point). Thus Neugebauer mentions (1957, pp. 31-32) the number 44,26,40 occurring in a table of reciprocals opposite 1,21—the latter number denoting $60 + 21 = 81$ in decimal notation. It is clear from the context here that the former number would be 0.0,44,26,40 if the decimal point and zero had been available—which they were not, since the tablet exhibiting the table was from the Old Babylonian period. Despite these deficiencies, however, the full use of sexagesimal fractions was available. As one might expect, only the *finite* sexagesimal fractions were comprehensible; in particular, the table of reciprocals just mentioned omitted reciprocals for numbers like 7 and 11, which "do not divide" 60 (approximations to such numbers are found in some texts, however). Nonetheless, the Old Babylonian scribes evidenced awareness of the fact that fractions could be treated like integers in computations through the device of expressing them in place value notation. And the existence of multiplication tables as well as tables of reciprocals showed that they made full use of this awareness. This stands in striking contrast to the Egyptian mathematics in which symbolization of fractions took a special form (see Section 2.4).

2.2 *Cipherization*   Further exemplification of the operation of cultural stress is to be observed in the improvement and simplification of numerals. The earliest forms of written number symbols were probably tally marks; thus |, ||, |||, |||| for 1, 2, 3, 4 respectively. These forms persisted in the Sumerian numerals for integers from 2 to 9, but because of the difficulty and time involved both in writing and in ascertaining the number of marks in a tally such as |||||||| or even in the more legible form $\begin{smallmatrix}||||\\||||\end{smallmatrix}$ sometimes used, it could be expected that changes and improvements would evolve. Generally, as cultures evolved into higher forms in which agriculture, commerce, and the like appeared, the double necessity of (1) using large numbers and (2) maintaining written records could be expected to force the introduction of improved ideographs.

Despite their potentiality for expressing symbolically all integers and fractions, the Babylonian numerals possessed a grave defect, from a practical point of view. This lay in the fact that the symbols for individual integers, although an improvement upon primitive tallying, were still complicated and cumbersome to handle. For instance, using the later cuneiform script in which a unit was represented by a wedge-shaped symbol ⟨Y⟩ and 10 by ⟨ , the number 48 might appear as

What was needed was further *symbolization;* more precisely, the invention of new symbols to stand for complicated groups of old symbols. This type of innovation was carried furthest by a culture

employing a number system that did not use the place value system at all, namely the Egyptian.[16]

In this culture, the device of substituting a single new symbol for a group of old number symbols indicating an integer was carried to a point where, for ordinary computations, the cumbersome features such as those that one encountered in the Babylonian numerals were eliminated. For the sake of brevity, this type of symbolization (of numerals) may be called *cipherization,* a term employed by Boyer (1944), who maintained that the process of cipherization deserves more emphasis in the history of numeral development than it has been accorded. In most ancient numeral systems, only halfway measures were taken in this direction. A common practice was to introduce a new symbol for 10, sometimes for 5 (as in the Mayan system). But intervening numbers were designated by combining the symbols so obtained with the symbol for the number 1; thus in the old Roman numeration, 9 was represented by VIIII or sometimes by IX (using a subtractive principle).[17] And although certain numbers such as 50 or 100 might also be given special symbols, the writing of larger numbers, even when (as in the Babylonian and Mayan systems) place value was introduced, was quite cumbersome.

Undoubtedly this was usually due in great measure to cultural lag; once a style for writing numbers is established in a culture, alterations take place very slowly. Special circumstances may intervene, of course; as Boyer suggests (1944), the use of a writing tool which (as in the Babylonian culture) is severely limited in the variety of possible marks that can be made with it, may act as a special cultural impediment to the invention of new symbols. On the other hand, the Egyptian tools constituted no such obstacle to

[16] So far as I have noticed, no one seems to have suggested the influence of cumbersome ciphers as a force toward invention of place value notation.

[17] The form IV for four, instead of IIII, was not used in antiquity, but is a modern extension of the subtractive principle; see Boyer (1944, footnote 24).

symbolic invention. The great advantages of more extensive cipherization were not realized in antiquity except in the Egyptian
hieratic script, and later, the demotic. In the hieratic, special
symbols were used for 1, 4, 5, 7 and 9, but 2 and 3, for instance,
were still denoted by the tally symbols || and |||, respectively. In
the demotic, however, *all* digits from 1 to 9 were assigned special
symbols (see Boyer, 1944, p. 157).

2.2a   *The Ionian numerals*   The numerals developed by
the Greeks provide an interesting example. Although the (older)
Greek Attic numerals were not well cipherized, the later Ionian
system was. To the 24 letters of the Greek alphabet were added
three archaic letters, and the resulting 27 letters were used as
follows: the first nine were assigned to represent, respectively, the
integers from 1 to 9; the next nine letters to the nine integral
multiples of 10 (i.e., 10, 20, 30, etc.); and the remaining nine
letters to the first nine integral multiples of 100 (i.e., 100, 200,
300, etc.). As a result, all numbers under 1000 could be assigned
numerals consisting of at most *three* simple symbols (the convention was to write the symbols from left to right in descending order
of rank). Later the multiples of 1000 were designated by the first
nine letters of the alphabet preceded by a stroke $(,\alpha,\ ,\beta,\ ,\gamma,\ \ldots)$,
and a new symbol, M, introduced for 10,000—the myriad—whose
multiples could be written as αM, βM, . . . (or, alternatively,
$\overset{\alpha}{M}, \overset{\beta}{M}, \ldots$).

The main weakness of the Ionian system was its inability to
symbolize larger and larger numbers indefinitely; for this, continually new symbolic devices would have to be invented. Nevertheless, the system is termed by Boyer (1944, p. 159) as "perhaps
the greatest single advance ever made in numeration and practical
arithmetic." For it not only served all the purposes for which the
counting numbers were needed at the time, but, as amply demonstrated by Tannery (who, by way of experiment, memorized the

system and calculated with it), was just as efficient for ordinary calculations as our own modern number system.[18]

For modern purposes, the lack of fractional representations comparable to our decimal fractions would be another, and fatal, weakness of the Ionian system, but for the Greeks, who were interested only in finite approximations, the Egyptian system of unit fractions sufficed. Their astronomers, however, realizing the effectiveness of the Babylonian place value system in this regard, adapted the latter to their uses, mixing the Ionian integral representations and the Babylonian fractions, precisely as we moderns do in writing degrees, minutes, and seconds in terms of "Hindu-Arabic" numerals (see below).

2.3 *Fusion of place value and cipherization* That the Babylonian numerals with their potentiality for expressing all integers and fractions were not adopted by the Greeks (or other cultures) is not surprising in view of their cumbersome symbols. The latter were not suited to quick and easy computation in the marketplace, and unless one was an astronomer the advantages of place value were not usually apparent. There is little doubt that acquaintance with the Babylonian system existed in Greece; but diffusion does not occur when adoption of a cultural element would offer no functional advantage, or is not forcibly imposed (*cf.* Preliminary Notions, Section 2). A more easily handled system was needed for the ordinary purposes of life, and this was afforded by the Ionian numerals. Although adequate proof of such a hypothesis seems not to exist, it appears not improbable that the inception of the Ionian system was influenced by the Egyptian manner of writing numerals.

The way was now open, however, for the introduction of a numerical system that incorporated the advantages of both place

---

[18] See the examples in Boyer (*loc. cit.,* p. 162) comparing multiplication in the Ionian and modern decimal systems as well as in the Babylonian and Egyptian hieroglyphic.

value and cipherization. To use a term that will be discussed more fully later, the opportunity for *consolidation* was now present. This is one of the tools that the modern mathematician frequently employs; when he observes two or more features of separate mathematical systems that seem to complement one another, or which would combine to form a more powerful or efficient theory, he will set about constructing such a theory by consolidating them. It is, of course, possible that certain individual mathematicians of the era under discussion became aware of the advantages of consolidating place value and cipherization. This appears to have been actually the case.

For scientific purposes where fractions were employed, as in astronomy, the advantages of the Babylonian place value system were realized; moreover, since the Greek astronomy followed the Babylonian tradition, it was also quite natural to continue with the Babylonian tabular system. But because of the cumbersome form of writing integers, the curious custom arose of using Ionian symbols for the individual digits (an analogous custom is followed today, as already noted, except that we now replace the Ionian symbols by our own). Neugebauer states (1957, p. 22): "Ptolemy . . . uses the sexagesimal place value system exclusively for fractions, but not for integers. Thus he will write 365 [in Ionian numerals] as τξε (300, 60, 5) but not as Ϛ ε (6, 5)." And this practice evidently was not peculiar to one astronomer. The Babylonian tablets of the Seleucid period (300-0 B.C.) show a remarkable development of astronomy, and the sexagesimal system that was used exclusively in the related calculations was passed on to the Greek astronomers and their followers. But the sexagesimal notation was "rarely applied with the strictness with which it appears in the cuneiform texts of the Seleucid period in Mesopotamia." And the practice of mixing different numeral systems "was followed by the Islamic astronomers and is the reason for our present astronomical custom to write integers decimally and then use sexagesimal minutes and seconds" (Neugebauer, 1957).

Only a partial consolidation of place value and efficient cipherization was achieved at this time, and indeed the type of consolidation needed, such as is exemplified in our modern numeral system, was not to become effective for many centuries (e.g., see Boyer, 1944, p. 164). To those who express surprise that the ingenuity of the Greek mathematicians did not produce a satisfactory consolidation, one can point out that "It is entirely possible that the Greeks were aware that, through the use of the principle of local [place] value and of a symbol for an empty position [zero], one could dispense with all but the first nine alphabetic numerals, but that they felt that there was little to be gained through the change. After all, would such a form as δoo be superior in any way to ν for 400, or would εοοοοοο be a significant improvement over φM? Let him raise his voice in criticism who has never either determined incorrectly the number of zeros in a product or made an error through placing a digit in the wrong column. As Tannery remarked, the alphabetic notation offers certain definite advantages, and the Greeks may well have regarded these as justifying its retention."[19] [Quoted with publisher's permission.]

2.3a *"Hindu-Arabic" numerals*   It would be nice to be able to sketch, at this point, a neat, orderly picture of the evolution of what today are termed inaccurately the "Hindu-Arabic" numerals. But unfortunately extant records are too meager, so that historians cannot agree on the details. There is fairly general agreement, however, that (1) the digits from 1 to 9 that we use today derive from Hindu forms, (2) the Hindus used a symbol for zero, at first a dot but later the oval form, and (3) they had, at least by A.D. 800, a positional notation for integers and made some use of negative numbers.

But just how much the Hindus borrowed from other cultures is a matter for conjecture and dispute. That Babylonian influence was a factor in the development of Indian mathematics is well

[19] Carl B. Boyer, "Fundamental Steps in the Development of Numeration," *Isis,* 1944, 35: 153-168, on pp. 164-165.

established, but to what extent cannot be definitely stated. For example, was the Indian zero a cultural descendant of the Babylonian zero? H. Freudenthal points out (1946, note 27) that during the period from A.D. 200 to 600, when the decimal system was coming into use in India, the Hindus became acquainted with

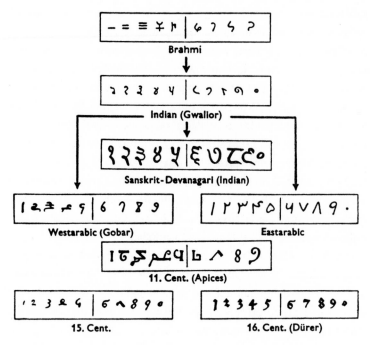

Genealogy of our digits. (From Menninger, Zahlwort and Ziffer, p. 329.)

Greek astronomy. And as a by-product of their interest in Greek astronomy, the Hindus also became acquainted with the sexagesimal place value system and the use of a symbol to indicate absence of a numeral (a "zero"). As additional evidence, Freudenthal points out that the so-called Hindu "versified numbers" always placed units first, tens next, and so on, while the Babylonians and Greeks used the reverse order. At the time when the Hindus commenced using digital symbols, the order conformed to that of the

Babylonians rather than to that of their native versified forms.[20]

At any rate, during the first few centuries of the post-Christian era the decimal place value system for *integers* seems to have developed in India, complete with zero symbol. Moreover, this zero symbol became a numeral in the *operational* sense, since it was used like any other numeral in operations such as addition and multiplication. That it became a numeral standing for a number *conceptually* is doubtful; standing alone, as a symbol, it probably did not. Its original invention was unquestionably merely to perfect the place value notation, standing solely as a symbol for "nothing"; it could have been suggested by the various forms of the abacus, which by this time were in rather general use in the East (from Rome to China). A defect of all forms of the abacus was its lack of a recording element; like the early computing machines, once a computation was performed, it was erased to make way for a new computation. If one wished to preserve the result of a computation, he had to make note of it in some form of numerals. And what would be more natural, given a numerical result on the abacus, such as 2301, where the column represented by the 0 is vacant, than to introduce a symbol corresponding to this empty column? Whatever the details of the invention might have been, it was undoubtedly a case of cultural stress acting on the symbolic level. But to achieve the conceptual status of number was to require further centuries of use.

The decline of the Hellenistic and Roman empires gave the chief role on the stage of history to the Arabic culture, to which the Hindu numeral system gradually diffused, taking its place alongside other systems of enumeration such as the Greek alphabetical system. There are indications that adoption of the Hindu numerals by the Arabs was promoted by a type of cultural resistance (*cf.* Preliminary Notions, Section 2) based on a prejudice against Greek culture. For example, Struik states (1948*b*, p. 87): "Our

[20] A readily accessible discussion of Freudenthal's conjecture may be found in B. L. van der Waerden (1961, p. 56).

first extant reference to the ten Hindu numerals outside of India is found in the writings of Bishop Severus Sebokht (662 A.D.), who mentions these numerals for the express purpose of showing that the Greeks did not have a monopoly of culture. [Struik makes reference here to Smith, 1923, Vol. I, pp. 166-167.] There existed under the early Abbasid khalifs in Bagdad a mathematical school which seems to have refused intentionally to accept lessons from the Greeks and turned for inspiration to the ancient Jewish and Babylonian sources. To this school belonged Al-Khowarizmi, the father of our algebra. The hatred of Greek influence and domination was one of the chief causes for the easy victories gained by the rising Islam in Syria." [Quoted with publisher's permission.]

From the Arabic culture the new decimal numerals finally diffused to European cultures by way of Spain and Italy, along scholarly and trade channels. The process was slow and gradual. The numerals underwent symbolic changes as they traveled[21] and encountered strong cultural resistance. An interesting example of the latter was a statute of 1299 forbidding the bankers of Florence to use the Arabic numerals and insisting on retention of the Roman ones. (See Struik, 1948a, Vol. I, p. 105). "Why then did the Italian merchants select the Arabic system? The answer to this question does not seem to have been given, but some facts may be suggestive. 'Arabic civilization was tangible enough in Sicily and Spain, not only in the Moorish part of it, but in places like Toledo, which were again in Christian hands; the Arabic speaking people encircled the Mediterranean Sea and dominated the Asiatic trade.' (Sarton, G., "Introduction to the History of Science," II, Baltimore, 1931, p. 6.) The Arabic system seems to have appeared in Italy first in Florence and Pisa which had closer connections with the Arabic world than with the Greek and which were in endless commercial rivalry with Venice" (Struik, 1948b, p. 88). [Quoted with publisher's permission.]

---

[21] See Struik (1948a, Vol. I, p. 88) for instance, and references therein.

2.4 *Decimal fractions*   The next step necessary in the evolution of the Hindu-Arabic numeral system was to extend it, as the Babylonians had learned to extend the sexagesimal system, to fractions.

The Egyptians used mainly "unit fractions," such as ½, ⅓, ¼, and so on, converting other fractions to sums of these. (An exception was ⅔, for which they had a special symbol.) For common uses, the Greek employed these Egyptian fractions, along with other systems. On the other hand, the Hindus used fractions that were similar to ours and were written like ours but without the bar; thus ¾ would be written $\frac{3}{4}$. (If written with an integer, as in 7 ¾, the integral part would top the array; thus $\frac{7}{3}$.) The utility of such special symbols for fractions is attested to by the fact that we still employ them in modern cultures. On the other hand, the advantages of a uniform system for writing whole numbers and fractions are obvious (see Section 2.1a). And once the Hindu-Arabic numeral system gained supremacy in Europe, it was inevitable that with the advance of science it would be extended to fractions.

As remarked above, the Babylonian sexagesimal system never died out, especially for writing "rational" fractions (Section 2.1a) in astronomical tables. At the same time, as might be expected the "obvious" (hindsight!) decimal analogues of these sexagesimal fractions began to appear here and there in isolated instances over the centuries. Neugebauer, commenting (1957, p. 23) on the "perfection to which Islamic scholars developed numerical methods" points out that it has recently been found that an astronomer named al-Kāshi, who died in 1429, invented the decimal analogues of sexagesimal fractions and stated approximations to the value of 2π in both sexagesimal and decimal fractions. Karpinski (1925, p. 123) states that Leonard of Pisa (c. 1170-1250; called "Fibonacci"—"son of Bonaccio"—as used in the term "Fibonacci series") gave an approximation to a root of a cubic equation in sexagesimal fractions carried to the eighth place and that during the Middle Ages such fractions were used in all computations. "In approximations of square root and cube root results were frequently given in sexagesimal fractions. However, Johannis de Muris in the fourteenth century gave the square root of 2 as $1 \cdot 4 \cdot 1 \cdot 4$, saying that the 1 represented units, the first 4 tenths, the second 1 'tenths of tenths,' and the second 4 'tenths of tenths of tenths.' However, he then extended this, writing the result also to twentieths of twentieths, finally giving the result in sexagesimal fractions" (Karpinski, 1925, pp. 128-130).[22] [Quoted with publisher's permission.]

The delay in the general use of decimal fractions appears to be a prime example of cultural lag. It was inevitable that someone would recognize the true situation and make an overt attempt to correct it. Stimulus for the breakthrough (see remarks in Chapter 2, Section 3.1a *re* "hereditary stress") to true decimal fractions was provided by such compilations as Regiomontanus' (1436-1476) tables of sines using a radius of 10,000,000 and cotangents with a radius of 100,000. "In both cases only a decimal point was necessary with a unit radius to give the modern tables" (Karpinski, 1925, p. 130). If complete details were to be filled in, the natural evolution of modern decimal fractions could be set forth showing the final eruption, in a conceptual sense, in the work of Simon Stevin in 1585,[23] who apparently gave the "first systematic discussion of decimal fractions with full appreciation of their significance" (Karpinski, 1925, p. 131). Stevin's notation was clumsy in that each digit was accompanied by a number in a circle indicating the decimal place—zero for units place, 1 for tens, and so on. For full acceptance, a better device was needed to indicate positional value, such as the decimal point or

[22] Quoted by permission of Rand McNally and Co., from L. C. Karpinski, 1925.

[23] For a survey of the history of decimal fractions, see Sarton, 1935.

an equivalent separation symbol. According to Karpinski (1925, p. 133), the decimal point appeared in print in 1616 in the English translation of Napier's logarithms. But to the end of the seventeenth century other writers used other notations such as the left half of a pair of brackets. Moreover, decimal fractions were still not universally adopted until the eighteenth century; until then, according to Karpinski (1925, p. 135), "arithmetics which avoided any mention of decimal fractions were about as numerous as those which gave some treatment of them."[24]

3   EVOLUTION OF THE CONCEPTUAL ASPECT OF NUMBER

In the discussion of the evolution of written numeral systems in Section 2, the conceptual aspect of number—roughly, the "meanings" of the numerals—was touched on here and there (e.g., in Section 2.3a). In the present section, this will constitute the main subject of inquiry.

When did number words first come to have an objective meaning? In grammatical terms, when did they become "nouns"? Clearly, the earliest uses of such words were adjectival—their descriptive character, even when such "nouns" as "ear" (for "two") or "man" (for "twenty") were used, was manifest in the manner of their use, as it was also in the manifold cases of their adaptations to various categories of objects (as in the cases cited earlier—see Section 1.2e).[25] But words change in both use and meaning. For instance, the English word "contact," originally a noun, has acquired verbal uses today. When the author was a student he could "make contact" with his teacher, but he certainly could *not* "contact" him—not if he hoped for an "A" grade. The

---

[24] It is interesting to note that Karpinski recognizes Stevin as "an independent discoverer of decimal fractions" (1925, p. 133) and later (p. 135) says that "it is perfectly evident that a succession of thinkers made possible this attainment. . . . In the field of science we are truly the heirs of all the ages past." In the latter connection he comments that "the development of decimal fractions illustrates the process of evolution in the realm of mathematical ideas."

[25] Extensive discussion of number words as adjectives will be found in Menninger (1957).

process can even change adjectives to verbs; the author recently heard a committee member, who was obviously tired of prolonged argument over minor details of the wording of a motion, suggest turning the problem over to the chairman with instructions to "vague it up" (so as to submerge the trifling differences).

For a word to become a noun, it must stand for something, and the process involves a change in conceptual status. In this way abstract concepts evolve, such as *good* and *evil*. And this ultimately happened with number words; as soon as the word "two" becomes a noun, it must stand for something, the concept of "twoness." The introduction of ideograms, special symbols for numbers (i.e., numerals), accelerated the process. But it did not affect all numbers simultaneously. For example, "zero" had no number status until centuries after its introduction—perhaps it never attained such a status in the Mayan culture, and certainly did not in the Babylonian.

3.1 *Number mysticism; numerology*   Like other sciences, mathematics had its mystical periods (and its mystical offshoots—but here only the main line of its development is considered). And although not all cultures necessarily practiced numerology, enough instances are on record to lead to the belief that the development was a natural stage in the evolution of number concepts. As a rule, only counting numbers were involved.

In the Babylonian culture, the influence of astrology on number concepts was probably strong. The following remarks concerning the status of the number 7 in Babylonian times are taken from a work on medieval number symbolism (Hopper, 1938, pp. 16-17):

"Having discovered 7 days, 7 winds, 7 gods and 7 devils, the astronomer proceeded to search for 7 planets and, what is remarkable, to find them! His quest was long and difficult. In early days, only Jupiter and Venus were recognized as planets. But when he had found 7, his task was ended; he need seek no further. The planets became the 'fate-deciding gods,' and at a much later time were appointed to rule the days of the week, a notion which seems not to have gained general currency until the first century B.C. at Alexandria.

"Meanwhile the Babylonian priest-geographer divides the earth into 7 zones, the architect builds Gudeas's temple, 'the house of the 7 divisions of the world,' of 7 steps. The Zikkurats, towers of Babel, originally 3 or 4 stories in height but never 5 or 6, were dedicated to the 7 planets and came to consist of 7 steps, faced with glazed bricks of the 7 colors, their angles facing the 4 cardinal points. These 7 steps symbolize the ascent to heaven, and a happy fate is promised the person who ascends to their summit. The tree of life, with 7 branches, each bearing 7 leaves, is perhaps the ancestor of the 7-branched candlestick of the Hebrews. Even the goddesses are called by 7 names and boast of them." [Quoted with publisher's permission.]

Whether such evidence justifies concluding that a number like 7 had achieved "noun" status can of course be debated, but it is difficult to believe that performance of a merely descriptive function would accord such status to 7. Perhaps to concede that mystical status lies somewhere between an ordinary descriptive character and the objective character of a noun, in a manner difficult for the modern mathematician to conceive, is to come near the truth of the matter. It is interesting to speculate, indeed, whether this stage in the evolution of number is not actually representative of the status of number in the popular mind of today! Probably the "average" modern "man in the street" has precisely the same conception of number (and perhaps of mathematics *in toto*) that the Old Babylonian had. Consider only the belief in "lucky numbers," for instance, or the prevailing faith in numerology and the persistent successful careers of numerologists (*cf.* Bell, 1933).

A classic instance, described by W. W. R. Ball, of the survival of the mystical power of the number 7 in the modern period is cited in Moritz (1914, p. 212): "The discovery [of Ceres] was made by G. Piazzi of Palermo; and it was the more interesting as its announcement occurred simultaneously with a publication by Hegel in which he severely criticized astronomers for not paying more attention to philosophy, a science, said he, which would at once have shown them that there could not possibly be more than seven planets, and a study of which would therefore have prevented an absurd waste of time in looking for what in the nature of things could never be found."[26]

3.2 *A number science*   As happened also in the case of the natural sciences, mathematics benefited from mysticism. This becomes unquestionably apparent in the Pythagorean school (see Section 3.4), but seems also to have been the case earlier in Babylon. Tablet after tablet of purely numerical work testifies to the affection that the temple scribes must have formed for their "number science." Recent researches into the achievements of Babylonian mathematics justify calling the mathematics of that age a "number science." Elaborate multiplication tables were constructed. It was recognized that division by a number n is equivalent to multiplying by its reciprocal $1/n,$ and remains have been uncovered of extensive tables for reciprocals to be used for division. (Ingenious methods were worked out for finding reciprocals.) Mathematical problems, some of them like the problems that one used to find in school textbooks, which could hardly have had any connection with an actual "real life" situation, testify to the abstract character of this number science. Moreover, it eventually evolved into "algebra," an algebra without the symbols that we usually associate with it—a "word" algebra. This included a standard method—quite modern except for the paucity of symbols —for solving quadratic equations. And even cubic and quartic equations (as well as higher-degree equations reducible to lower degrees) seem to have been solved. Tables of powers were evi-

[26] From Robert E. Moritz, *On Mathematics and Mathematicians,* Dover Publications, New York, 1914. Reprinted by permission of the publisher.

dently used for solving, by interpolation, the type of exponential equations encountered in ascertaining the term required for money to grow to a given amount at a given interest rate.

That most of these skills were developed under cultural stress seems clear. Evidence exists of the love of "pure" arithmetic, with no idea of application; but generally the problems solved, even when rather removed from physical "reality," bear an unmistakable stamp of the influence thereof. As in the case of Egyptian mathematics, we find problem after problem worked out, but little of what a modern mathematician would call "theory."

Two further aspects of this Babylonian science of numbers might be noted here, particularly because of their significance for later developments in mathematics. One of these concerns the concepts of theorem and proof. Historians usually point out that Babylonian mathematics contained nothing that we would call a theorem, that is, a general statement with logical proof thereof. It was a "do this, do that" kind of affair. [Here the analogy to much of the teaching of elementary mathematics as it was practiced in today's schools prior to the "reforms" (see Introduction, Section 4) makes one wonder if it had progressed beyond the Babylonian stage.] Clay tablets contain abundant testimony to the "drill" that the prospective scribe went through, working problem after problem all of the same type. Nonetheless, we find evidence in this practice of what we might call "intuition" of a theorem; that no explicit statement of the theorem is given does not mean that a general property or rule that a modern mathematician would embody in a theorem was not known. In particular, the statement that the Babylonians knew the Pythagorean theorem (see Chapter 2, Section 2) is not meant to imply that an explicit formulation of it has been found, but only that the tables of "Pythagorean numbers" and other evidence indicate an awareness of the content of the theorem.[27] Similarly, their processes for solving equations, such

---

[27] See especially Neugebauer (1957, pp. 35-36). Concerning a tablet on which the length of the diagonal of a square is determined from an ap-

as simultaneous systems and quadratics, although exemplified only by example after example, might very well have been stated in suitable symbolic "theorem" form if the requisite algebraic apparatus had been available. They apparently were in a stage where progress in symbolic representation of their abstractions had not developed to the point of explicit presentation in theorem form. They clearly knew, however, that certain procedures would always achieve certain results, even though they might not state these as general propositions.

Furthermore, recent researches of Neugebauer and Sachs have revealed that the Babylonians did give proof of an elementary sort for some procedures. Tablets recently deciphered show that they made use of their knowledge of reciprocals to prove—that is, "check"—the reciprocals in their tables. The number n was recorded, together with the key numbers used in finding $1/n$; then $1/n$ was handled in the same way, to show that the reciprocal of $1/n$ was n. From a broad viewpoint, one can consider such methods to be precursors of the later, more refined concept of "proof." More likely, however, the prevailing "proofs" were pragmatic—the demonstration that a "theorem" (here only a verbally stated procedure or formula) "worked" in example after example; this accumulation of examples constituted the "proof." Even today this is the kind of validation of a theory that one expects in the natural sciences, and it might be inferred from the Babylonian records that it was the type of proof used in Babylonian mathematics. After all, "proof" is only a form of *verification,* and in present-day mathematics not every mathematical statement is verified by logical methods. Usually it is only the general theorem that we prove logically. For the special cases of the Babylonian mathe-

---

proximation to $\sqrt{2}$ to three sexagesimal places, Neugebauer remarks: "The above example of the determination of the diagonal of the square from its side is sufficient proof that the 'Pythagorean' theorem was known more than a thousand years before Pythagoras." This tablet is shown opposite p. 90.

matician, verification needed no logic. (We emulate him when we "prove"—that is, verify—that 2 is a root of the equation $x^2 - 3x + 2 = 0$.)

To summarize this aspect, then, of Babylonian mathematics, it is possible that their mode of communicating their methods was by verbal statement of rules, leaving to posterity only the examples they used to illustrate these rules, baked on their clay tablets. And if such was the case, they were not far from the concept of a general theorem. To object that a "theorem" without logical proof is a long way from being truly a theorem would be a judgment on the basis of our own standards; we must not forget that what constitutes "proof" varies from culture to culture, as well as from age to age. Their type of "proof" may not have been so reliable or refined as that of the later Greeks (whose standards in turn were hardly up to our own, incidentally), but it was quite valid in their eyes and quite satisfactory from the viewpoint of the standards of their own culture.

3.3 *Status of the number concept and its symbolization at the end of the Babylonian ascendancy*   Where did the Babylonians leave the number *concept?* That they made a remarkable advance is now recognized. From what is known of the beginnings of the Mesopotamian culture that is designated by the label "Babylonian" it can be inferred that its "mathematics" commenced with elementary forms of counting, based variously on bases of 10 and 60 or mixtures thereof. So far as the numerals for natural numbers (the "counting" numbers) are concerned, the Babylonians ultimately attained (except for cumbersome cipherization) in the sexagesimal symbolism a system as adequate as that in use today, complete with zero and place value, enabling one to symbolize arbitrarily large numbers. Moreover, they recognized that the same system could be extended to symbolize fractions. They stopped short of its full applicability to arbitrary fractions, since this would have necessitated the notion of an infinite array of symbols. However, not even their brilliant successors in the

Greek world, nor, for that matter, the entire civilized world up to the nineteenth century, reached such a concept or realized its importance for the theory of the general real number (see Chapter 3). The latter development required a type of cultural stress internal to mathematics (to be described later) that was present, to be sure, in the Greek era but which resulted at that time only in the geometric forms of number known as "magnitudes" and in the brilliant work of Eudoxus.

That the Babylonians arrived at some kind of number *concept* is evidenced by the apparent affection the scribes felt for "numerology" (as in the case of the number 7, for instance). And there seems little doubt that the Babylonian number science, although probably restricted to special practitioners (such as the temple scribes), was at least of as high a cultural level as that represented by the number knowledge of the average college graduate of today. Moreover, this number science evidently constituted in the Babylonian culture the entire element that would today be called "mathematics." The geometry occurring therein appears to have been no more than a subsidiary element (see Chapter 2), of importance for its practical value, as well as a fertile field for the application of the number science; and there is also evidence of a methodological nature reminding one of the Pythagorean uses of geometric forms in number theory.

3.4 *The "Pythagorean" school*   Contemporary with the later developments in Babylonian science, there flourished in the Greek culture the group of philosophers and mystics known as the "Pythagorean school." Tradition has it that this "school" was founded by one Pythagoras who lived "about 572-497 B.C. or a little later" according to Heath (1921, Vol. I, p. 67) and is supposed to have been a pupil of Thales. Most accounts relate that he spent most of his earlier years traveling and assimilating the mathematical and astronomical ideas of the Egyptians and Babylonians, finally settling in a Greek seaport (Crotona) in southern Italy and there founding his school. Although Pythagoras, as perhaps also Thales, may have existed only as a means for focusing in a special individual the tradition concerning certain origins, there can be little doubt about the existence of such an early group, having religious and political overtones, which developed a curious amalgam of number and geometry. With this group, number probably attained the most mystical and absolute character that it has enjoyed before or since. "Everything is number" is the phrase ordinarily employed to encapsule the Pythagorean philosophy. "Numbers were immutable and eternal, like the heavenly bodies; numbers were intelligible; the science of numbers was the key to the universe" (Russell, 1937, pp. ix-x). The establishing of a correlation between numbers and harmonics is often cited as the genesis of the philosophy.[28]

Unfortunately, the origins of the Pythagorean concepts seem likely to remain one of the great mysteries of all time. Much less is known about them than about the origins of the Babylonian number science; in the latter case clay tablets have survived, documenting much of the early nature and uses of numbers, while none of the writings of the Pythagoreans remain. Instead, reliance must be placed on later writers to whom Pythagoras was already a legendary figure. It is not even clear whether Aristotle believed in his existence or not, nor whether Aristotle was settled in his own mind regarding the Pythagorean concept of number (see, e.g., Heath, 1921, Vol. I, p. 66). Appeal to personal traits of individuals such as Thales and Pythagoras (who may possibly have been invented only as "explanations") is no explanation at all from a cultural standpoint. By way of contrast, the origins of Newton's work, for example, are quite clear from our knowledge of the state of mathematics at the time in which he worked.[29] But of the cultural milieu in which the Pythagorean ideas were formed

---

[28] That is, harmonious sounds are produced by strings with lengths in ratios expressible by the natural numbers.

[29] In Newton's own often-quoted words, "I stand on the shoulders of giants."

we know virtually nothing. Reliance on later writers would be
rash, in view of the disagreements among them and the obviously
fictional elements scattered through their accounts. We do know
that a prominent characteristic of the Greek philosophy was its
questioning and inquiring curiosity regarding the nature of man
and the universe and that emphasis was placed on formulating
"first principles." That in such an atmosphere a "school" with
mystical leanings should arrive at a theory of existence based on
number is perhaps not surprising.

But aside from this, one may justifiably be curious as to why
more of the Babylonian number science did not affect Greek
thought by way of diffusion. Here, again, one can only conjecture.
Such effects are clearly discernible in the later Hellenistic culture,
especially in the work of Ptolemy (the astronomer) and Diophantus. But in the period c. 700-500 B.C., during which the Pythagorean ideas were probably developing, culturologically speaking,
there seems to be little evidence of Babylonian influence. There
was possibly awareness of Babylonian number systems among
Greek philosophers without perception of the possibilities inherent
therein. Before a deliberate selection would be made of the sexagesimal place value notation from amongst the welter of already
existing (Greek and other) number systems, a clear need (cultural stress) would have had to be operative (as there was, later,
in Hellenistic astronomy). Presumably such a need was not existent, especially at the time (c. 600 B.C.) of which we are speaking and, moreover, at a time when the sexagesimal notation may
not even have included a zero symbol (aside from such other
defects as were mentioned in Section 2.2).

That the Greek culture was amenable to change is clear from
the fact that the earlier Attic numerals were gradually replaced by
the later Ionian alphabetical system (see Section 2.2; also Heath,
1921, Vol. I, pp. 30 ff). And, as already observed, the later Greek
astronomers did adopt the Babylonian sexagesimal number system.
But there is little evidence that it would have had any general

appeal during the early period of which we are speaking. With our hindsight, knowing the possibilities of the Babylonian numerals, this may seem an unfortunate circumstance.

The characteristics of the Pythagorean number mysticism have been treated in great detail in various works. For our purposes we need recall only a few of these at this point (further details will be found in Chapter 2). Just as the Babylonians "associated with each of their gods a number up to 60, the number indicating the rank of the god in the heavenly hierarchy," so did the Pythagoreans attribute "divine significance to most numbers up to fifty. . . . Even numbers [2, 4, 6, . . .] they regarded . . . as feminine, pertaining to the earthly; odd numbers as . . . masculine, partaking of celestial nature" (Dantzig, 1954, pp. 40-41).

"Each number was identified with some human attribute. *One* stood for reason, because it was unchangeable; *two* for opinion, *four* for justice, because it was the first perfect square, the product of equals; *five* for marriage, because it was the union of the first feminine and the first masculine number [2 + 3 = 5]. (One was regarded not as an odd number, but rather as the *source* of all numbers)" (Dantzig, 1954, pp. 41-42).

This mysticism also yielded benefits to mathematics. For the Pythagorean preoccupation with number apparently led to what is today called number theory (going much further than the meager beginnings of the Babylonians mentioned earlier). The Greeks called it arithmetic (by which name it has until recently been denoted in Europe). It is tempting to conjecture (which is all one can do) how a theory of genuine scientific interest arose along with, and from, such a profound number mysticism. One can argue that the numerological properties of numbers elevated them to an importance not attained in their utilitarian stage of development; from "lowly" origin they had now become of mystical significance. What more natural, then, than that their inherent properties be investigated, particularly since these might lead to hitherto unrecognized possibilities of mystical interpretation?

The distinction between evenness and oddness was of mystical significance; but in addition, from a number-theoretic point of view, this classification is basic. The notion of "amicable" or "friendly" numbers, later of interest to Fermat, Descartes, and Euler (Heath—1921, Vol. I, p. 75, footnote 1—remarks that Euler described sixty-one pairs of amicable numbers), has also been attributed to the Pythagorean school. Two numbers are amicable if each is the sum of the proper divisors of the other, and folklore has it that Pythagoras on being asked "what is a friend"? replied "One who is the other I, such are 220 and 284" (See Dantzig, 1954, p. 45; also Heath, 1921, Vol. I, p. 75). "This pair of numbers attained a mystical aura, and superstition later maintained that two talismans bearing these numbers would seal perfect friendship between the wearers. The numbers came to play an important role in magic, sorcery, astrology, and the casting of horoscopes" (Eves, 1953, p. 55).[30] Similarly, the "perfect" number, a number that is the sum of its proper divisors,[31] is credited by some as having been of interest to the Pythagoreans [although Heath (1956, Vol. II, p. 294) maintains that they used the word in a different sense]. Perfect numbers have held great number-theoretic interest from Greek times to the present. A classical result, found in Euclid's *Elements* (see Heath 1956, Vol. II, p. 421, Proposition 36), is that if $2^n - 1$ is a prime number (where n is a natural number greater than 1), then the number $2^{n-1} \cdot (2^n - 1)$ is perfect. Thus for $n = 2$, the number 6, the smallest perfect number, is obtained. Since such numbers are obviously even, one wonders if all perfect numbers are even—an unsolved question in number theory [Euler showed that if a perfect number is even, it must have the form stated in Euclid's theorem (Eves, 1953, p. 56)].

Whatever the factual historical details, it seems fairly clear

[30] Quoted by permission of The Macmillan Co.

[31] For example, 6 is a perfect number, since its proper divisors are 1, 2, and 3, and $6 = 1 + 2 + 3$.

that number mysticism played an important role both in the maturing of the concept of number and in the inception of number theory. Numerology and number theory ultimately parted, much as astrology and astronomy parted (and as, later, alchemy and chemistry took different paths), the one remaining as part of the stock-in-trade of the modern "numerologist" and the other (to quote Gauss) becoming the "queen of mathematics." Presumably, this splitting had not occurred in the Pythagorean school, the two being welded there by a common bond. And with "Pythagoreanism," number (meaning natural number—1, 2, 3, etc.) probably achieved a degree of abstractness equal to that of modern times.

4 INTERLUDE The evolution of the number concept, from its inception in primitive elementary forms of counting to the Greek idealization of number, involved an increasingly abstract type of thinking. Moreover, it was a process implemented by cultural forces already cited: environmental and particularly cultural stress, which induced primitive counting; symbolization, which not only helped number to achieve both an objective ("noun") and an operational status, but provided the "time-binding" vehicle through which it could continue its growth; and diffusion between cultures situated in the Mesopotamian plain and later continued throughout the Hellenic area—all these contributed to the growing abstraction of the number concept, as, indeed, did consolidation (Section 2.3). Even cultural lag and cultural resistance played a not inconsiderable part in the process. By the time of the Pythagoreans, counting and ideograms for numbers had diffused throughout the eastern Mediterranean area, and the number idea reached a maturity that found expression in the abstractions of Pythagorean mysticism. The culmination of this evolution, usually associated with the name of Plato, was the reification of number to a domain of ideals in which reside the "true" forms not only of numbers, but of other mathematical concepts. From this point of view, numbers are existent independently of their human uses, both preceding the advent of man and presumably after his eventual disappearance from the earth. This conception of number is still widely held by the layman as well as by numerologists, philosophers, and mathematicians. Its philosophical validity or non-validity poses no problem for us; however, its existence as a cultural phenomenon does concern us, as do all the varying conceptions of number that have evolved.

Between the end of the Greek era and the eighteenth century, little progress was made in the evolution of number. The Romans, being a "practical" people, were little interested in mathematics other than what was required in computation, and they had learned to use the abacus for this.[32] This fact stands in startling contrast to the Greek love of theory as opposed to practice. Although the "Hindu-Arabic" numerals underwent some minor symbolic alterations during the interim between the time of the Roman Empire and the European Renaissance, their general character underwent little significant change.

With development of the complete decimal number system (including decimal fractions) during and after Stevin's time (Section 2.4) further progress could be expected. Little advance was made, however, until the "experimentalisms" of the seventeenth, eighteenth, and early nineteenth-century analysts were completed. By this time it became clear that the meaning of "real number continuum" needed to be given an explicit and clear foundation.[33] This was undertaken during the latter half of the nineteenth century and early part of the twentieth. And since geometry was destined to play a part in this, the account of it will be delayed in order to review the early evolution of geometry, especially with regard for its relation to number.

---

[32] It is interesting to cogitate regarding the influence of the abacus on the development of number. Although it may have been instrumental, in earlier times, in suggesting place value and the invention of a zero symbol, it may later have acted as an impediment to the invention of further improvements in numeral systems.

[33] It is one of the "miracles" of mathematical history that so much "good" analysis was created without such a foundation. However, the hereditary stress furnished by this work was precisely what was needed to force the invention of a foundation.

CHAPTER TWO

Evolution of Geometry

> If arithmetic had remained free from all admixture of geometry, it would have known only the whole number; it is to adapt itself to the needs of geometry that it invented anything else. Poincaré (1946, p. 442)

1 THE POSITION OF GEOMETRY IN MATHEMATICS   Probably the "average layman," when confronted with the word "mathematics," thinks immediately of computation, that is, operations with numbers—the sort of thing that the Babylonians would have considered mathematics to be (if they had had such a word). On the other hand, anyone who went through high school (at least prior to the modern revisions in the curriculum; *cf.* Introduction, Section 4) probably remembers that among the courses offered in mathematics was one called "Geometry," or "Plane Geometry." If he took this course, he found that it was radically different from the type of mathematics that he had taken before. And this despite the fact that in both his arithmetic and his algebra courses he had undoubtedly worked problems having to do with mensuration of geometric figures. In the geometry course, however, he found that he started with some "assumptions" called "axioms" or "postulates," along with some "definitions," and proved "theorems," "corollaries," "lemmas," and so on, by methods called "logical." The whole process of development of the subject was entirely unlike that by which he became acquainted with number operations and algebra.[1] Perhaps he wondered why such was the case, and at first, if he was an "unusual" child, he may have pondered over

---

[1] Recent changes in teaching methods have introduced the so-called axiomatic method into arithmetic and algebra, however, as well as made the change from the latter to the courses in geometry much less abrupt.

why this new type of subject was labeled mathematics in the first place. If so, he would not have been alone in his cogitations; consider the quotations that we give below.

The first is from a book by a well-known analyst.[2] An analyst —one who works in analysis—bases his ideas on numbers, not geometry, although analysis owes a great deal to ideas suggested by geometric situations, and modern analysis makes considerable use of a branch of mathematics called topology whose origin was geometrical. One of the most famous theorems in topology is called Jordan's theorem, or Jordan curve theorem, and the analyst whom we are quoting found it necessary to take account of it:

"Jordan's theorem is . . . unavoidable in proving that *mathematics does not involve geometry,* Euclidean or otherwise" (Italics mine: R.L.W.).

This is a curious statement in that it invokes a geometrical theorem to expunge geometry from "mathematics." And now contrast it with the following, taken from a book[3] on the foundations of geometry:

"Any objective definition of geometry would probably include *the whole of mathematics*" (Italics mine: R.L.W.).

These statements, so contrasting in their viewpoints, were not made by eccentrics, but by eminent and respected mathematicians (at least two of whom, Veblen and Whitehead, made fundamental contributions to mathematics by their researches). This is not the proper place to discuss why such opposing views could have been held by mathematicians contemporary with one another; we shall be in a better position to do so later. For the present, they may be considered as evidence that professional mathematicians are themselves not agreed on what constitutes geometry, any more

[2] Paul Dienes, *The Taylor Series; an Introduction to the Theory of Functions of a Complex Variable,*" Oxford, Clarendon Press, 1931. The quotation is from the Preface, p. v.

[3] O. Veblen and J. H. C. Whitehead, *The Foundations of Differential Geometry,* Cambridge Tracts in Mathematics and Mathematical Physics, No. 29, Cambridge University Press, 1932; see p. 17, footnote.

than they are agreed on what constitutes mathematics. Moreover, they suggest the question, *how did geometry ever get into mathematics in the first place?*

Despite these observations, however, there are some objects of study that every mathematician would call geometry. An outstanding example is that of the geometry of Euclid. Presumably Paul Dienes, to whom is due the first quotation above, would not have dignified this geometry by assigning it a place in the category of mathematics. And for this he had some justification, it must be admitted. It is not uncommon for one to hear the remark that Euclidean geometry is really part of physics. And if one reflects on the fact that the Greeks believed that their geometry established a science of space and spatial relations—where by "space" they meant physical space—there is certainly good ground for this assertion. The word itself—geo-metry—is a compound of Greek words meaning "earth" and "measure"; in other words, "geometry" literally stands for "earth measure," so that from etymological considerations, one would think that a branch of civil engineering was indicated by the word. Moreover, after the introduction of analytic geometry by Descartes and Fermat in the seventeenth century, one could argue that the necessity of the Euclidean system could no longer be substantiated.

But words constantly change or take on new meanings, and this has happened to the word "geometry." Also, one now has non-Euclidean geometry, projective geometry, differential geometry, and so on, as well as subjects like topology that, although they have apparently broken out of their original geometric confines, still embody a great deal of material that most mathematicians would label geometry. Whether one would agree that these subjects, in all their manifestations, form a part of *mathematics,* is another matter; one may insist that only when they are couched in the language of algebra and analysis are they acceptable as mathematics. On the other hand, one may oppose to this demand the opinion that only the fundamental concepts are of importance;

whether they are dressed up as analysis or are stated in the language of ordinary discourse is immaterial, so long as no distortion
of the concepts takes place. It may be only a matter of taste how
one wants to present his ideas; the important thing is to present
them clearly. Most mathematicians abhor presentations that are
cumbrous and difficult to manipulate. Thus some theorems of
Euclidean geometry are easier to prove by algebraic methods than
by the classical synthetic (logical) methods; but the converse is
also true, and consequently probably most mathematicians accept
both methods and thereby, by implication, choose to retain geometry in a position of respect in the mathematical world.

But to return to the changing character of the meaning of
the word "geometry": Just as we do not hesitate to use the word
"mathematics" despite the fact that it is hardly an exaggeration
to say that no two mathematicians could agree on a definition of
the word, so we do not hesitate to use the word "geometry." It
has a general connotation that is useful despite disagreements over
the propriety of the various special applications of the word. Actually, most present-day use is modified by accompanying the word
with a suitable adjective, as in the terms "descriptive geometry,"
"algebraic geometry," and others already mentioned. But aside
from this one frequently hears a man labeled a geometer, implying
that he works in a field known as geometry; just as one sometimes
hears a department chairman remark that he would like to have
someone who could teach his courses in "geometry."

2 PRE-GREEK "GEOMETRY"   Now, it was not always
thus. There was a time when *mathematics* included nothing that
one would place in a separate category and label geometry. The
present-day "intuitionist," who considers mathematics as what can
be derived by constructive methods from the natural numbers (see
Chapter 5, Section 2.2), would have been quite happy in that
early environment. For at that time mathematics consisted solely
of an arithmetic of whole numbers and fractions, together with an
embryonic (albeit quite remarkable) algebra—hampered, to be

sure, by symbolic shortcomings that would drive any modern mathematician, intuitionist or not, to distraction. Recent investigations have established that this early mathematics reached a quite astonishing level, in spite of symbolic handicaps. However (as already pointed out in Chapter 1, Section 3.2), of *proof*—an integral part of modern mathematics—there is no evidence in this ancient mathematics, except for such elementary forms of proof as may be inferred from the "checking" of results (e.g., proving the correctness of a reciprocal by showing that its reciprocal is again the original number). It is wrong, however, to say that the Babylonians did not prove their rules (see Chapter 1, Section 3.2). But proofs were empirical, as in the natural sciences today. Not surprisingly, this resulted in some of their rules being incorrect. Nevertheless, methodologically the Babylonian algebra was quite remarkable, giving not only solutions of quadratic, but of higher-degree equations as well. Especially noteworthy is the investigation of Pythagorean triples, established by the so-called Plimpton 322 tablet.[4] Moreover, even the solution of exponential equations was carried out, as in the determination of the time required for a sum of money to accumulate to a desired amount at a given rate of interest; this was accomplished by the use of tables, as one might expect. To quote Neugebauer (1957, p. 48): "However incomplete our present knowledge of Babylonian mathematics may be, so much is established beyond any doubt; we are dealing with a level of mathematical development which can in many aspects be compared with the mathematics, say, of the early Renaissance."

But what of "geometry" during this period? Naturally, one would not expect to find what we would today call geometry. But without abusing the use of the word at any stage of its develop-

---

[4] See, for instance, O. Neugebauer (1957, pp. 36 ff). Chapter 2 of this book is an authoritative as well as popular description of the present state of our knowledge of Babylonian mathematics. Evidently the Babylonians thought of the "Pythagorean" relation as numerical—a relation between numbers, while for the Greeks it was a *geometrical* relation—a relation between areas.

ment through a variety of meanings, one can find little that one would call geometry in the Babylonian mathematics. This statement requires some explanation, especially if the Babylonians were aware of the "Pythagorean theorem" over a thousand years before Pythagoras.[5] Actually, one can amplify the statement by adding that neither was there anything in his mathematics that a Babylonian would have called geometry, if by this is meant a term signifying a special branch of mathematics. The relation between the sides of a right triangle embodied in the theorem of Pythagoras was no more or less than any other rule of mensuration known at the time; no more than a mathematical rule for measuring the amount of money in a fund at interest for a certain length of time, for instance. The Babylonian would no more have considered that the Pythagorean rule was part of a special branch of mathematics called by a special name than a modern mathematician would consider that the "mathematics of finance" was a special branch of mathematics; in the latter case, only a special set of techniques forming an application of mathematics is indicated, rather than a branch of mathematics proper. Similarly, the Babylonian mathematician would have considered rules for finding areas and the like as only a special set of techniques forming an application of his number science.

He had worked out methods for finding areas of both plane and solid figures (not always leading to accurate results). (See, e.g., the list in Archibald, 1949, p. 8). As Neugebauer put it (1957, p. 45), "At this level there is no essential difference between the division of a sum of money according to certain rules and the division of a field of given size into, say, parts of equal area. In all cases exterior conditions have to be observed, in one case the conditions of the inheritance, in another case the rules for

---

[5] For a list of some geometric "theorems" known to the Babylonians, see Archibald (1949, p. 8). These include, for example, rules for finding areas of rectangular right triangles, certain trapezoids, and volumes of prisms, right circular cylinders, and the frustrum of a cone or square pyramid.

the determination of an area, or the relations between measures or the customs concerning wages. The mathematical importance of a problem lies in its arithmetical solution; 'geometry' is only one among many subjects of practical life to which arithmetical procedures may be applied . . . 'geometry' is no special mathematical discipline [in Babylonia] but is treated on an equal level with any other form of numerical relation between practical objects. These facts must be clearly kept in mind if we nevertheless speak about geometrical knowledge in Babylonian mathematics, simply because these special facts were eventually destined to play a decisive role in mathematical development." That is, *we* speak of "geometrical" items in Babylonian mathematics simply because with our knowledge of subsequent developments based on these items we can single them out for special attention.

So far as Egyptian "geometry" is concerned, a rule for finding the area of any triangle seemed to be known, but no convincing evidence of an acquaintance with the Pythagorean theorem. On the other hand, it is remarkable that it included a formula for the volume of a frustrum of a square pyramid which is essentially that now in use. (See Sarton, 1959, Vol. I, pp. 39-40).

3   WHY DID GEOMETRY BECOME PART OF MATHEMATICS?
In view of these facts, it is natural to ask, "How did what we call geometry ever get into mathematics at all?" If the other special applications of mathematics in the Babylonian sense, such as the applications to finance, astronomy, and so on, never became part of "mathematics" in the subsequent development of mathematics in Western culture, why did the material that was concerned with measuring areas of fields, volumes of casks, and the like, manage to creep into the mathematical fold? Why did this material not become a part of "engineering," for example? Of course it did become part of the *equipment* of every engineer and physicist, but the point is that it became much more—even part of what we today call pure mathematics. (Presumably Dienes, whose remark anent mathematics not including geometry was quoted above,

would admit that it is a part of mathematics when clothed in suitable analytic dress.)

The answer, of course, can be found only by searching through history. Unfortunately, history at this point becomes quite hazy, and we are suddenly confronted with the spectacle of a rapidly developing Greek mathematics having a character highly geometric both in content and method and contemporaneous with the later development of the Babylonian numerical mathematics. Perhaps due to the destruction, sometimes careless and wanton, of early manuscripts and clay tablets, we have to rely for our factual historical information on writers to whom it apparently never occurred that there was anything like what some today call the Greek miracle. Evidently what occurred must have seemed quite natural to the participants in the mathematical revolution that produced Greek geometry.

If any answer is to be found, then, we must assemble what facts there are available and try to construct a theory, with indulgence in as little guessing as possible. Of course, one can use a "cutting the Gordian knot" procedure and attribute the whole affair to "revelation," particularly since some of the earliest participants in the Greek developments were the mystical group reputedly founded by Pythagoras (see Chapter 1, Section 3.4); one then has truly a "miracle" as the answer. But this reminds one of the manner in which fossils were at one time explained as due to "fossil-making forces." And if one presumes that mathematics is just as much the result of cultural evolution as any other human construct, then one must use what knowledge he has of the manner in which cultures act and react to one another in order to form a reasonable explanation. Moreover, one will not expect to find a *single* component for the answer, such as "revelation," but a complex of interacting components, some internal to mathematics, some only partially so, and some quite external to the mainstream of mathematical evolution.

Conventional history inclines in such instances to look for

some *individual* to whom the "miracle" can be attributed. Plato has Socrates (in the *Phaidros*) say: "I heard, then, that at Naucratis, in Egypt, was one of the ancient gods of that country, the one whose sacred bird is called the ibis, and the name of the god himself was Theuth (Thoth). He it was who invented numbers and arithmetic and geometry and astronomy, also draughts and dice, and, most important of all, letters." No modern scholar would accept such an explanation, of course, but many do accept the stories told about Thales (c. 624-547) by such Greek historians as Herodotus and Proclus. He is singled out as the founder of geometry, and to him they credit such fundamental theorems as "A circle is bisected by a diameter," "The base angles of an isosceles triangle are equal," and "If two lines intersect, then opposite angles are equal" (*cf.* Archibald, 1949, p. 17). But Neugebauer (1957, p. 148) remarks that "This story clearly reflects the attitude of a much more advanced period when it had become clear that facts of this type require a proof before they can be utilized for subsequent theorems. To the later mathematicians it seemed natural to assume that theorems which had to be established first on logical grounds should also come first chronologically. Actually the Greek historians acted in exactly the same way as modern historians do when no source material is available to them; they restored the sequence of events according to the requirements of the theory of their own times. We know today that all the factual mathematical knowledge which is ascribed to the early Greek philosophers was known many centuries before, though without the accompanying evidence of any formal method which the mathematicians of the fourth century [B.C.] would have called a proof."

In many ways we are in a better position today to explain the inception of Greek geometry than were the historians closer to the events concerned. From recent researches much is known about Egyptian and Babylonian mathematics that was not within the perspective of Greek historians, apparently. Moreover, there is

more awareness of the processes of cultural evolution, even though
not always agreement on how they operate—enough so, at any
rate, that instead of looking for miracles or gods or superhuman
individuals, ways may be sought in which the ancient Babylonian
and Egyptian *ideas* passed into the Greek culture. There can be
little doubt, for instance, that there was a continuous line from
the Babylonian and Egyptian mathematics to the Greek mathematics, and that although the Greeks may have borrowed many
geometric rules from Egypt, they probably borrowed their initial
awareness of the contributions that geometrical concepts can make
to arithmetic and algebra from the Babylonians.

3.1 *Number and geometric magnitude*   An early, and
basic, component in the assimilation of geometric concepts by
mathematics was evidently related to the use of numbers in mensuration, particularly in the measurement of length.[6] Indeed, there
can be little doubt that this association of *number,* the most basic
element of mathematics, with *line,* one of the most basic elements
of geometric form (the more basic element, *point,* was a later and
more mature concept), lay at the heart of the absorption of geometric form into mathematics. And even this simple notion was
the subject of long evolution; the early Greeks did not consider
that every line must have a length, for instance. But the germ of
the concept was there, emphasized and reemphasized every time
an ancient surveyor took out his measuring instruments or, and
more to the point, every time an ancient mathematician con-
cocted a problem based on the surveyor's action. Since it was the
Babylonian, as well as the Egyptian, custom to present mathematical concepts by means of suitably devised problems, geometrical
situations presented a prime source of such. Numbers were associated with area and volume; and since numbers as such are

---

[6] The origin of measurement and the creation of standards therefor is
analogous to the origin of number. In both cases direct comparison of
objects measured or counted preceded the abstraction to standards of length
or number. For an interesting account, see Childe (1948, pp. 193 ff).

combinable, the numbers associated with area and length were quite frequently indiscriminately combined by addition or multiplication. (The error inherent in such operations was not recognized until much later.)

3.1a *Geometric number theory* Also influential unquestionably was the Pythagorean number theory (see Chapter 1, Section 3.4), which associated number with geometric form in devious ways. The classification of numbers as "triangular," "square," "pentagonal," and so on, is a good example. With a pointer the triangular numbers were represented with dots in the following manner:

$$
\begin{array}{ccc}
& & \bullet \\
\bullet & & \bullet \; \bullet \\
\bullet \quad \bullet & \bullet \; \bullet \; \bullet & \\
1 & 3 & 6
\end{array}
\quad \text{and so on.}
$$

Square numbers were similarly represented by squares:

$$
\begin{array}{ccc}
& \bullet \quad \bullet & \bullet \; \bullet \; \bullet \\
\bullet & & \bullet \; \bullet \; \bullet \\
& \bullet \quad \bullet & \bullet \; \bullet \; \bullet \\
1 & 4 & 9
\end{array}
\quad \text{and so on.}
$$

But geometrical forms were also used to derive number-theoretic facts. For instance, by superimposing the representations of square numbers on one another

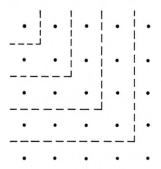

and by noticing that each superimposition could be accomplished simply by bordering the preceding square with an odd number of dots, the conclusion was inescapably reached that the sum of any number of consecutive odd integers, starting with 1, is a square; that is,

$$1 + 3 + 5 + \cdots + (2n - 1) = n^2$$

a formula that today is usually proved either by the rule for summing an arithmetic progression or by mathematical induction.

So far as purely geometric matters are concerned, it is problematical just how much the Pythagoreans accomplished. It seems probable that the "rules" known to the Babylonians and Egyptians were inherited by the Pythagoreans. Some say that Pythagoras discovered the theorem named for him quite independently, but many incline to doubt this. Perhaps he devised a proof for it and thus established credit in the same manner that the modern mathematician wins renown by giving a proof of what was formerly only a conjecture. Mention has already been made in Chapter 1, Footnote 21, of a tablet (in the Yale Babylonian collection) from Old Babylonian times on which is clearly impressed a square with its diagonals, accompanied by numbers indicating how to compute the length of the diagonal of the square from that of the side by using an approximation to $\sqrt{2}$ which, in decimals, is 1.414213 (the correct value being 1.414214 . . .). Not only is this tablet one of the indications that the Babylonians knew the Pythagorean theorem long before Pythagoras, but it is interesting for its association, at this early period, of number (in the sense of "real number") with line.

A concomitant, and possibly a consequence, of the Pythagorean "everything is number" philosophy was the intrusion into geometry of the assumption that all line segments are commensurable; that is, the lengths of two line segments have a ratio like that of one natural number to another (see Section 3 of the Introduction). Expressed symbolically, if L_1 and L_2 are the lengths of the line segments, expressed in terms of a suitable common unit (such as centimeters or inches), then $L_1/L_2 = m/n$ where m and n are natural numbers. Such numbers are today called positive rational;

An Old-Babylonian tablet showing a square and its diagonals. On the horizontal diagonal can be discerned the numerals 1, 24, 51, and 10 standing for an approximation to $\sqrt{2}$. (Courtesy of the Babylonian Collection, Sterling Memorial Library, Yale University.)

in modern mathematics any number that can be expressed in the form m/n, where m is an integer (positive, negative, or zero) and n a natural number is called *rational;* and the symbol m/n is called a rational fraction. Thus a natural number is also rational—2, for instance, may be expressed as 2/1. Likewise, negative integers are rational—for example, -3 may be expressed as $-3/1$. And zero is rational since it may be represented by 0/1. All "proper fractions" such as 1/3, 8/9, and so on, represent rational numbers. Using this terminology, then, the Pythagoreans originally believed that all ratios of line segments are rational. This would not have been too serious, perhaps, except that they based their proof methods on the assumption; and when they ultimately discovered (historians seem agreed on this) that the ratio of the diagonal of a square to a side is not rational,[7] the effects were presumably traumatic. Nevertheless, the resulting "crisis" was highly beneficial to mathematics. Although folklore has it that the Pythagorean sect endeavored to keep the discovery a secret, and executed by drowning the member who eventually disclosed the knowledge, the truth probably is that it was only one of the stresses that impressed on the Greek mathematical world the necessity for reconstructing and improving both the foundation and proof methods of geometry. (Another stress was provided by the Zeno paradoxes concerning the continuity of space and time.) As is well known, the Greek philosophers continually sought for basic elements on which to build their theories—in mathematics by logical deduction—and if a basic assumption was recognized as unjustified, presumably its correction became a problem of current research. In the present case the "correction" is credited to a non-Pythagorean, Eudoxus, called by Archibald (1949, p. 20) "an original genius second only to Archimedes." His treatment of incommensurables and proportion is generally considered as furnishing the content of Book V of Euclid's *Elements,* as well as being a natural forerunner of the nineteenth-century Dedekind theory of real numbers (see Chapter 3). In any event, the *continuous,* in the form of the continuous geometric line of magnitudes, was evidently forcing its way into mathematics, and the imbedding of the concept "number" in the ordered continuum of

---

[7] This ratio is $\sqrt{2}$; the usual proof of its irrationality will be found in footnote 2 of Chapter 3.

real numbers (not perfected until the latter half of the nineteenth century) was already under way.

The importance of this "crisis," from an evolutionary point of view, is that it furnishes the first clear-cut example of the action of "internal" cultural stress in the evolution of mathematics, a force that I prefer to call "hereditary stress."[8] This is a cultural stress created by the accumulation over a period, usually of extended duration, of concepts and their interactions *within* a system. It is frequently involved, as in the case being discussed, in the production of "crises."[9] In the present case developments that compelled recognition of an unsuspected type of number forced attention on finding a possible solution. No outside—that is, environmental—stress was responsible. The existence of hereditary stresses has been noted by historians of science; for example, Sarton (1952) remarked: ". . . each scientific question suggests irresistibly new questions connected with it by no bounds but the bounds of logic. Each new discovery exerts as it were a pressure in a new direction. . . . The whole fabric of science seems thus to be growing like a tree; in both cases the dependence upon the environment is obvious enough, yet the main cause of growth— the growth pressure, the urge to grow—is *inside* the tree, not outside [Italics ours]." Clearly by "growth pressure" Sarton meant hereditary stress. And in modern mathematics, where the "logic" is so manifest, the action of hereditary stress is conceivably more influential than in the natural sciences.

Perhaps if the Greek mathematicians had at this point a thorough awareness of the potentialities of the place value system as used in the Babylonian sexagesimal system, especially for representing fractions, they might have made a start on "modern"

[8] The stress is termed "hereditary" rather than just "internal" in order to emphasize that it was derived from earlier mathematical concepts that had combined to produce the stress.

[9] Compare Kuhn, 1962.

analysis. Instead, they continued in a geometric direction. This was natural in that the Pythagoreans had already begun their investigations in number theory by representing numbers by line segments or, as in the case of "square," "triangular," and so on, numbers, by higher dimensional geometric configurations, and also because the "crisis" regarding incommensurable numbers such as $\sqrt{2}$ had arisen in the *geometric* representation. An advantage, too, of the Eudoxian treatment of the theory of proportion was that the representation of numbers by line segments afforded a treatment of irrationals on the same basis as rationals.

3.2   *Number theory in Euclid; number and magnitude*
A curious (for us) dichotomy occurs in the Greek development as a result of the geometric treatment of number. For illustration of this we may again draw upon the *Elements*. First, a distinction is made between *number* and *magnitude*. As the term is used in Euclid, "number" means "natural number" in our terminology, with the exception of 1, which is termed a "unit" and is that of which a "number" is composed (*cf.* Heath, 1956, Vol. II, p. 277). On the other hand, "magnitude" is a concept like that of the extent or length of a geometric segment, a proper part of a segment being of lesser magnitude than the entire segment (although "magnitude" also included other geometric measures such as angle).

Book VII of the *Elements* is of special interest, particularly from the point of view of its probable pre-Eudoxian character, in that it develops elements of what we would call "number theory" (the Greeks called it "arithmetic" and used the word "logistics" to denote the use of numbers in the ordinary affairs of living) by geometrical methods. Book V of the *Elements,* embodying the Eudoxian theory of proportion, treats of magnitude; but then in Book VII the theory of proportion of "numbers" is treated independently. For example, Proposition 16 of Book V (Heath 1956, Vol. II, p. 164) states: "If four magnitudes be proportional, they will also be proportional alternately." (That is, if $a:b = c:d,$

then $a:c = b:d$.) And Proposition 13 of Book VII (Heath, 1956, Vol. II, p. 313) states "If four numbers be proportional, they will also be proportional alternately."

Another illustration concerns the well-known Euclidean algorithm for finding the greatest common divisor. For "numbers" this is treated in Proposition 2 of Book VII (Heath, 1956, Vol. II, p. 298): "Given two numbers not prime to one another, to find their greatest common measure." And for "magnitudes," Proposition 3 of Book X (Heath, 1956, Vol. III, p. 20) states: "Given two commensurable magnitudes, to find their greatest common measure."

Such repetition might be explained as due to a desire to treat the separate "books" as separate units; Heath accepts this explanation concerning the repetition just stated (Heath, 1926, Vol. III, p. 22). But regarding the repetition of the theory of proportion in Books V and VII he remarks (Heath, 1926, Vol. II, p. 113): "The latter exposition referring only to commensurables may be taken to represent fairly the theory of proportions at the stage which it had reached before the great extension of it made by Eudoxus. . . . why did Euclid not save himself so much repetition and treat numbers merely as a particular case of magnitude, referring back to the corresponding more general propositions of Book V instead of proving the same propositions over and over again for numbers? It could not have escaped him that numbers fall under the conception of magnitude. . . . Yet Euclid says nothing to connect the two theories of proportion even when he comes in X.5 to a proportion two terms of which are magnitudes and two are numbers ('commensurable magnitudes have to one another the ratio which a number has to a number'). The probable explanation . . . is that Euclid simply followed tradition and gave the two theories as he found them." If this explanation is correct, it was not to be the only case in mathematical history where tradition was to play such a strong role. Number, in the sense of natural number, held a special position in the evolution of Greek mathematics, and its name, "arithmos," had a quite different connotation from the term "melikotes," ordinarily translated as "magnitude."

It is of interest also to note that the term "ratio," whose Greek form was "logos," is given a "definition" in Book V—"A ratio is a sort of relation in respect of size between two magnitudes of the same kind"—but none in Book VII where ratios of (natural) numbers are studied. Possibly "ratio" of numbers was on enough of a "universal" basis—"common knowledge"—to need no explanation, but where incommensurables were concerned, something was needed to justify extension of the term to the new usage.

On the other hand, one may conjecture that the dichotomy is evidence of a feeling for categorization that the modern mathematician can well appreciate, namely, the separation of the number concept between its counting aspects (natural number) and its measuring aspects (the arbitrary real number). The natural numbers were used by the Babylonians and the Egyptians to ascertain the number of unit lengths contained in the entire extent of a physical object, and perhaps because of this practice their use in measurement was deemed quite as natural as their use in counting money. It was the Greeks who recognized the essential difference in the two roles of number, made manifest perhaps by the necessity for solving the incommensurability problem. It is a distinction like that which we preserve today in the notions of "cardinal number" and "ordinal number" (see Chapter 1, Section 1.2b), as well as in the differences between the natural number as "natural number" and as "real number." The role of the natural number as magnitude evolved so gradually in pre-Greek cultures that the distinction that the Greeks saw the necessity of making may have been obscured. And although, as Heath surmises, Euclid was following tradition in the separate treatment accorded natural number as magnitude, the logical justification for so doing is something that appears quite modern in nature.

The well-known theorem concerning the "infinity" of primes, which is embodied in Proposition 20 of Book IX of the *Elements* of Euclid, is an illustration of the Greek use of magnitude in number theory. Actually, what Euclid did was to give what we now call an *algorithm* (i.e., a rule for *construction* of some entity) showing how, given any finite collection of prime numbers, one can find a prime number not in the collection.[10] His geometric proof has, however, an exact numerical analogue: If $p_1, p_2, \ldots, p_k$ are the given primes, consider the number $n = (p_1 \cdot p_2 \cdot \ldots \cdot p_k) + 1$, and if n is prime, observe that it is a prime not among the given primes $p_1, p_2, \ldots, p_k$; if n is not prime, a prime factor of n will not be one of the given primes, since from the form $(p_1 \cdot p_2 \cdot \ldots \cdot p_k) + 1$ it follows that none of the given primes is a factor of n.

It must not be concluded, from the use of geometry to treat numbers as exemplified in the *Elements,* that every time a citizen of Athens wished to buy a new tunic, he had to get out ruler and compass to compute the cost. For as we have already observed (Chapter 1, Section 2.2a) the Greeks possessed a numeral system quite adequate for ordinary affairs and business transactions—the Ionian system. During the time when number theory was being investigated on the basis exemplified in the *Elements,* this "alphabetical" system of writing numbers (or its predecessor, the Attic system) was being used for ordinary computations—by the scholar as well as by the trader. Moreover, scientists such as Archimedes used it in their computations. The facility with which it could be used was perhaps partly responsible for the fact that it persisted in the East Roman Empire until the fifteenth century. In particular, it seems to have been much better suited to ordinary uses than the clumsier Roman numerals. It is interesting to recall that (Chapter 1, Section 2.2a) the French mathematician Tannery familiarized himself with the Greek numerals, practicing the four elementary operations in doing the calculations contained in Archimedes'

[10] His proof supposes three primes in the given collection, but it is obvious that the same procedure will handle any finite number of primes.

Measurement of a Circle. According to Heath (1921, Vol. I, p. 38), he found that it "had practical advantages which he had hardly suspected before, and that the operations took little longer with Greek than with modern numerals."

The remarkable thing is that the new element, geometry, so completely took over mathematics, at least methodologically, in the Greek culture.[11] Presumably the elevation of geometry to a dominant position resulted chiefly from the manner in which geometry could be used to discover and prove number-theoretic theorems, as well as to cope successfully with the general real number in the guise of "magnitude." Of course, it would be false to say that mathematics was entirely geometry to the Greeks— many of the later Greeks, such as Diophantus, clearly stood as much in the Babylonian algebraic tradition as in the Greek geometric one. And as pointed out by Felix Klein (1939, p. 193), Euclid did not intend the *Elements* to be an encyclopedia of all contemporary Greek mathematics; not even his (Euclid's) own work on conics was included. Nevertheless, in the absence of a suitable algebraic symbolism, even Archimedes resorted to geometrical symbols. And certainly by the end of the Greek era, mathematics had adopted geometry (if it had not been adopted by geometry!). For a time, Greek mathematics also adopted other subjects, such as music, but it is not difficult to see why these did not remain in the main body of mathematics, while geometry did. These other subjects were more in the nature of particular applications of mathematics, while geometry, being concerned more with abstract form, was obviously not restricted to any one sort of natural phenomenon.

3.3 *Concept of form in number and geometry*   This brings us to a final comment on the reasons for the adoption of

[11] The Pythagorean *quadrivium* consisting of arithmetic, geometry, music and astronomy, which seems to place arithmetic on an equal plane with geometry, persisted into the Middle Ages, however; along with the *trivium* of grammar, logic, and rhetoric it constituted the knowledge basic to a proper education.

geometry into mathematics. Just as in the case of the "geometrical algebra" to be found in Euclid,[12] in which geometry offered a representation of numbers and operations therewith of easy visual comprehension, so did geometry become a useful tool for the application of mathematics wherever *form* is central to the application. Astronomy is a good case in point; the Greeks conceived the paths of the planets to be circles and carried forward the study of astronomical phenomena on this basis. Now, number itself, especially the natural numbers, 1, 2, 3, . . . , is essentially related to the forms of *sets;* whether a collection is a singleton, a pair, a triple, and so on, is certainly basic to its form. If the set is ordered, this is an additional feature of its form; but number (i.e., cardinal number; see Chapter 1, Section 1.2b) is more primary. In this sense, geometry is an extension of the way in which we study form and patterns in mathematics. Probably this aspect of the matter also influenced the Greeks to such uses of geometry as they made in their treatment of arithmetic. Not that the Greeks had arrived at any such advanced *conception* of mathematics as to consider it a study of form or pattern, but they were forced to the realization in their treatment of the irrational or the incommensurable.

4 LATER DEVELOPMENTS IN GEOMETRY   Unfortunately, like most of the writings of antiquity, a major part of the Greek mathematical works was lost or destroyed. For our information about them, we have to rely principally on commentators who lived centuries later. Of Euclid's work we are fortunate in having, in nearly complete form, five of the at least ten works that he wrote; one of these is, of course, his famous *Elements*. This work has been considered ever since as the epitome of logical perfection (which it is not, however, by modern standards), in that from a few basic assumptions ("axioms" and "postulates") and a collec-

---

[12] Neugebauer sees the possibility of a direct connection between the geometrical algebra of the Greeks and the Babylonian solution of second-degree problems in which one was asked to find two numbers x and y from a knowledge of their product and of their sum or difference. (Neugebauer, 1957, pp. 149-150.)

tion of definitions, 465 propositions are deduced in a logical chain. As has already been observed, neither Babylonian nor Egyptian mathematics had arrived at such a standard of presentation; and just how it evolved in the Greek mathematics is a matter for conjecture and debate. (See Szabo, 1960, for instance.) Many surmise that the criticisms of Zeno and the discovery of incommensurables constituted a cultural stress sufficient to compel the search for a consistent *foundation* from which the scattered propositions of geometry, number theory, and algebra (such as it was) could be deduced. Euclid's *Elements* is undoubtedly the culmination of earlier works of a similar format and was so successful that it has remained almost to the present day the standard text for geometry of the plane and space (its number theory and algebra having, generally speaking, been replaced by more modern symbolic treatments). Its mode of logical deduction was so much considered the ideal that Archimedes and, nearly 2000 years later, Newton, both presented the finished forms of their masterpieces in a similar logical pattern (although each had a different method for originally deriving their results).

To follow the later evolution of geometry very far would require too many technicalities. During the period between the Greek era and the seventeenth century there was little evidence of new forms evolving. However, sufficient stress was generated by advances in symbolizing algebra, as well as advances in art, architecture, astronomy, engineering, and science generally, to effect the formation of new conceptual patterns in geometry. Only two of these developments will be mentioned here.

4.1 *Non-Euclidean geometry*   One of the fundamental assumptions on which Euclid founded the *Elements* was the parallel postulate—frequently called the parallel axiom.[13] In Section 3

---

[13] In the *Elements,* the fundamental assumptions are given in two groups, one called axioms, the other postulates. The axioms were evidently intended to be of a "universal" nature, such as "Things equal to the same thing are equal to one another." The postulates were the *geometric* assumptions, such as "All right angles are equal to one another."

of the Introduction the problem of whether this postulate could be derived, logically, from the others was described, and the motive for the years of work spent on the problem was attributed to "aesthetics." No mathematician, or any other scientist for that matter, likes to set down a collection of fundamental assumptions ("axioms") for a theory and include one that is logically deducible from the others.

In the case of the parallel postulate, this feeling ultimately took on all the features of hereditary stress, in the form of an almost compulsive challenge to prove the postulate from the others. There apparently was a kind of cultural "intuition," common to the entire mathematical community of the later Middle Ages, that the postulate could not possibly be "independent" of the others. Apparently it was felt the "nature of things" compelled this to be the case (see the account of Saccheri's work in the Introduction).

Now, it is fairly well known today that the solution of an outstanding problem is most likely to be accomplished by several mathematicians working independently, rather than by a single investigator. There is no plausible explanation for this phenomenon except on a cultural basis. The tools needed, the concepts suggesting the suitable analogies, and so on, all accumulate within the culture, being equally available to and assimilable by all workers in the field. And when their accumulation is sufficient to create the requisite stress, the problem commences to be solved—not by one, but by several investigators. Of course the solutions may not —and usually are not—exactly simultaneous; this would hardly be expected. But they cluster sufficiently close in time for simultaneous publication (as in the case of Bolyai and Lobachevski) to occur. Moreover, the number of those who achieve solutions and do not get to publish, or who are about to achieve them, usually remains unknown (Gauss was an exception in that his solution of the problem became known; that of a mathematician of lesser fame and status would not have). Although these facts are certainly familiar to most scientists, who can usually cite other cases personally known to them, the solution of the parallel postulate problem is of added interest because of the long period preceding the solution. One might feel that the longer the period over which a problem has been known and worked upon, the less likely would simultaneous solutions occur. But this overlooks the way in which cultural items evolve; in particular, the necessity for the requisite accumulation of new tools and concepts to occur. And in the case of the parallel postulate, the ideas beginning to take shape during the few years that preceded its solution, particularly concerning the formal character of axiomatic systems in algebra, undoubtedly furnished the new insights that led to the solutions.

4.2 *Analytic geometry*   The introduction of analytic geometry furnishes an excellent example of what was termed "consolidation" in Chapter 1, Section 2.3. It is an interesting fact that while the number theory and the geometrical algebra of the *Elements* were superseded and replaced by the symbolic advances of the early Middle Ages, the purely geometric aspects of the *Elements* continued to survive in their so-called synthetic form—that is, in the form of proof based on pure logic.

Nevertheless, the idea of introducing symbolic methods that would allow such shortcuts as had been achieved in algebra began to crop up by the seventeenth century. As might be expected, several innovators brought out such ideas virtually simultaneously—notably Descartes, Desargues, and Fermat. Without going into details, the general idea was to invent a device for representing the configurations of geometry by algebraic equations which could then be manipulated according to the rules of algebra to produce results that, interpreted geometrically, would constitute theorems important to geometry. In this way, the algebra that, in Greek mathematics, depended on geometry for its development and that (thanks to new symbolic methods) had grown independently to attain a new maturity, could now in turn apply to geometry in ways that turned out generally to achieve solutions of geometric problems much more simply than the classical logical methods.

What happened here, viewed from the standpoint of the seventeenth century, was that consolidation of two elements of the current mathematical culture, namely algebra and geometry, produced a new and more powerful mathematical method, *analytic geometry*.[14]

5 EFFECTS OF THE DIFFUSION OF GEOMETRIC MODES THROUGHOUT MATHEMATICS   What effects can be discerned in mathematics as a result of the "invasion" of mathematics by geometry? And were there beneficial *contributions* of geometry to the subject that adopted it? We have already seen some of the effects during the Hellenic period; these were of so fundamental a nature that not only was "mathematics" growing beyond the confines of a number science (as conceived in Babylonia), but Greek mathematics became essentially geometric. There were other decidedly beneficial contributions, moreover, that have had a profound influence on modern mathematics.

5.1 *Axiomatic method; introduction of logic*   In the first place one should probably put the invention of the *axiomatic method*. It was presumably in the attempt to develop geometry on a secure foundation, safe from such paradoxes as those of Zeno, as well as to provide for incommensurability, that the Greeks developed the axiomatic method in mathematics. In fact, our forebears came to think of axioms as *part* of geometry, just as students of recent times came to think of logarithms as part of trigonometry. It was in *geometry* that one used axioms, not in arithmetic or algebra (except as these might be incorporated into geometry themselves, as of course they were by the Greeks). So geometry must be credited for the introduction of the axiomatic method.

Interestingly enough, however, mathematicians seem to have

---

14 Some maintain (*cf.* Coolidge, 1963, pp. 117 ff) that analytic geometry began with the Greeks, who used a geometrical algebra that only needs to be interpreted in modern algebraic symbolism to give modern analytic geometry. The material above should not be considered a counterargument to this; it merely uses the bringing together of the newer algebraic symbolism and the geometry as an example of consolidation.

been among the last to adopt the axiomatic method as a *general* foundational device, that is, apart from geometry, since not until the twentieth century did its use as a basis on which to found systems other than geometric become very common. In contrast to this, during the seventeenth and eighteenth centuries there was a veritable deluge of attempts to develop social and philosophical theories—particularly ethical and political—on a postulational basis. The classic example is Spinoza's *Ethics;* although names closely identified with mathematics such as Descartes and Leibniz are also associated with such projects (as a young man Leibniz used the "geometrical method"—as the axiomatic method was then termed—to present solutions of political questions).[15] It was not until the nineteenth century that the axiomatic method commenced to be generally accepted in mathematics. Then the possibilities of the use of axioms not only as a means for founding and generalizing mathematical and physical concepts, but also as a research tool were revealed by such pioneering efforts as those of Hamilton, Gauss, Peacock, and others in algebra, Whewell in mechanics, Grassmann, Pasch, Hilbert, and others in geometry, and the Italian logician Peano and his followers in general formal systems. That it took so long for such an important theoretical tool to diffuse from geometry to the rest of mathematics was undoubtedly due to both cultural lag and cultural resistance.

As an integral part of the axiomatic method, *logic* also attained a more prominent status in mathematics. As a peculiarly Greek mode of thought, logic lay at the very heart of the axiomatic method. And it need hardly be pointed out what the ascendancy of logic in mathematics has meant to mathematics. Moreover, its importance in methods of proof is so great that ultimately some came to insist that mathematics is really an *extension* of logic, maintaining that the *essence* of mathematics is logical deduction. The "definitions" of mathematics attributed to the late Harvard mathe-

---

[15] For an interesting discussion, see Bredvold, 1951.

matician Benjamin Peirce ("Mathematics is the science which draws necessary conclusions," 1881), A. N. Whitehead ("Mathematics in its widest signification is the development of all types of formal, necessary, deductive reasoning," 1898), and Bertrand Russell ("Pure Mathematics is the class of all propositions of the form '*p* implies *q*,' where *p* and *q* are propositions . . . ," 1903), are typical of the conclusions reached by a large majority of the mathematical community around the turn of the century. However, it seems safe to say that this point of view has not many supporters today. In the meanwhile, applications of the axiomatic method to logic itself resulted in a so-called mathematical logic, as well as in exposure of the fact that, when analyzed axiomatically, "logic" is revealed as not really a unique theory—much as the uniqueness of Euclidean geometry was destroyed by invention of the non-Euclidean geometries. This mathematical logic furnishes another example of a theory that, pursued only in the interests of pure science (compare Section 3 of the Introduction), eventually turns out to have important applications (as in computer theory, for instance). Clearly, the effect of the ultimate widespread diffusion, throughout mathematics, of the logical methods of the Greeks has had a profound effect on mathematics and its applications.

5.2 *Revolution in mathematical thought*   Another effect of geometry on mathematics may be observed by considering its role in the revolution in mathematical and philosophical thought that occurred in the nineteenth century. This revolution, to be sure, can be attributed in large part to the increasing use of the axiomatic method in algebra and in formal logic. But the introduction of the non-Euclidean geometries gave a decided stimulus to the movement. As a result of this revolution it became evident that mathematics was not bound to certain *special* patterns, given a priori as Kant imagined, or patterns found in our perception of the external world, but that it could create *its own patterns,* limited only by the current state of mathematical thought and the significance that such patterns may have for mathematics or its applications. Without this freedom of the mathematical imagination to roam uninhibited by special applications, modern mathematics could hardly have been born. And for this mathematics has to thank geometry in large part. Moreover, the new freedom spilled over into other branches of science, especially physics, one of whose foremost leaders, Einstein, acknowledged his debt to the realization of the conventional nature of axioms that had evolved in pure mathematics.

5.3 *Effects on analysis*[16]   One can of course develop analysis without geometry. But it seems generally conceded that for the neophyte in analysis the geometrical representation of function and derivative, Argand diagrams for complex numbers,[17] and so on, are a great aid to comprehension. And the same is true of the evolution of these notions; the history of the early development of what is now called classical analysis shows how much the analysts depended on geometrical concepts as both a creative and expository device. Indeed, some early analysis approached the state of the Greek geometrical algebra, so fundamental did the position of geometrical representation therein become.

In the centuries preceding the work of Cauchy and his successors, the geometric concepts of curve and tangent dominated the development of the calculus. The notion of the integral as a limit of areas and that of the derivative as the slope of the tangent to a curve are still found pedagogically useful in the teaching of calculus. True, the calculus did not find a foundation rigorous enough to satisfy the conscientious mathematician until the geo-

---

[16] The reader unfamiliar with elementary forms of analysis, such as calculus, may omit Section 5.3 without loss of continuity with what follows.

[17] A Norwegian surveyor, C. Wessel (1745-1818), invented the so-called Argand diagram prior to Argand, but did not receive credit since his work was published in a journal not ordinarily read by mathematicians. (*Cf.* Bell, 1945, p. 177, for instance.) Moreover, Wessel wrote in Danish, so that his audience was very restricted.

metric wrappings were discarded in favor of the purely arithmetic treatments based on the concept of the continuum of real numbers, as developed by the late-nineteenth-century mathematicians (Weierstrass, Dedekind, and others). But from an evolutionary standpoint, the contributions of geometry to the development of the calculus were fundamental. One cannot say that they were *necessary,* since a different path from the notion of number to the calculus might conceivably have been taken. But the point is that the latter did not happen, any more than man bypassed earlier forms of life in his evolution; and quite possibly the geometric concepts *were* necessary to the natural evolution of analysis (a statement with which many present-day pedagogues of evolutionary persuasion seem to agree). For a revealing and scholarly history of this evolution, describing the struggle between the arithmetical and the geometric, the reader is strongly advised to read Boyer, 1949.

However, a much greater contribution to analysis, as well as to algebra, was and is being made by a type of geometry that has now burst its geometric bonds, namely *topology.* It may itself be considered as one of the contributions of geometry to mathematics, and its applications to other fields may therefore be properly included as one of the benefits that mathematics has received from geometry.

5.4 *Labels and modes of thought*   At this point it is well to recall the quotation from Veblen and Whitehead with which we began this chapter, namely that today "Any objective definition of geometry would probably include the whole of mathematics." In a way, it is largely a matter of *words.* Words inevitably carry their conventional connotations, and when one speaks, for instance, of "the linear continuum" he may mean the Euclidean line and denote its elements as "points" or he may be thinking of the "real numbers." In the former case, he is probably thinking of the "x-axis" of analytic geometry; in the latter case, he is thinking of the structure arrived at by starting with the natural numbers and building up the real number continuum. Which one he prefers is a matter of taste.

Certain parts of mathematics are labeled "geometry"—as in Euclidean geometry, algebraic geometry, differential geometry— just as certain parts are labeled "analysis" and other parts "algebra." But these labels again seem to be a matter of words and their conventions. The author is reminded of a letter from a former student who was studying at the Institute for Advanced Study at a time when certain aspects of group theory that had evolved in topology were being incorporated into modern algebra; he stated: "When I hear the word 'algebraic' being used in a mathematical discussion, I find that the users are usually topologists; and when I hear the word 'cohomology' and like terms derived from topology, I usually find that the users are algebraists."

Although labels are convenient and quite useful in their proper context, they should not be allowed to conceal underlying facts. And the fact is that it would be quite impossible to conceive of a modern mathematics in which the geometric, or what has derived from the geometric, was deleted. It would seem proper to interpret the statement of Dienes that "mathematics does not involve geometry" as only an indication that he preferred not to use *geometric terminology* as well as not to *think in geometric patterns*. That some people think visually and others not has been widely observed; and granting the distinction, perhaps it is the case that Dienes was one who did not think visually, so that geometric patterns were of little value to him, whereas Veblen and Whitehead, both of whom made notable contributions to geometry and its derivatives (such as topology), are to be numbered among those who thought visually. The author has long had a feeling that he could classify the algebraists with whom he has been intimately associated into the two groups, visual thinkers and nonvisual thinkers; partly because, being a "visual thinker" himself, he seemed to understand the former much more easily and partly because the manner in which they present their concepts reveals an underlying geometric pattern. Fortunately modern mathematics can accommodate both types; one can surmise that a nonvisual thinking Greek of 300 B.C. who had the mental capacity to become a reasonably good algebraist would never have dreamed of his potentiality, since the mathematics of his culture was so imbedded in the geometric mode of thought.

But one must not conclude that the predominance of geometric methods implied that all Greeks thought visually. There simply was no algebraic symbolism available for them to use, while the geometric patterns available were capable of symbolizing algebraic relations, such as sums, powers, square roots, and so on, and consequently it was these geometric patterns that were developed. To say that the Greeks made a "wrong turn" in this is to be blind to the fact that the Greeks could hardly do otherwise than use the tools of a symbolic nature that were predominant in their culture. To be sure, they could have used the Babylonian number system, and did to some extent, but the Babylonians did not have any *algebraic* symbolism to bequeath to the Greeks. So the Greeks used their own, albeit more cumbersome, number system and represented algebraic operations by geometric symbols—the so-called geometrical algebra. They had no other way to turn. And neither did their successors until the final evolution of algebraic and analytic symbols during the period beginning with the work of Vieta (see Struik, 1948*a,* pp. 115-118).

But the thesis should not be belabored too much. Probably enough has already been said to indicate that the effects on mathematics, and in particular on the evolution of number, made by the diffusion of geometry and geometric modes of thought throughout mathematics have not been inconsiderable; indeed, it would be quite impossible to imagine what mathematics would be like without geometry. It has contributed symbolically, conceptually, and psychologically to the evolution of mathematics. Moreover, the Greek geometry, far from being a wrong turning in the evolution of mathematics, was a natural development from the cultural elements existing at the time it was created and quite as necessary, perhaps, as the prehuman primate was to the evolution of *homo sapiens*.

CHAPTER THREE

The Real Numbers.
Conquest of the Infinite

Because of the predominance of geometric modes of thought in later Greek mathematics, and the reintroduction of these ideas into medieval Europe, the development of mathematics in the later Middle Ages and the Renaissance was patterned, both conceptually and to a great extent symbolically, along geometric lines. But it was inevitable that with the concurrent growth of physical theories—and in those days the mathematician and physicist were most likely one and the same person—the geometric modes that were so theoretically useful in analysis would have to be replaced by an equivalent numerical pattern. The problem of incommensurables that the Greeks had so deftly sidestepped through the device of substituting magnitude for number, now had to be faced anew. So long as numbers were used for measure, "magnitudes" might suffice. But when the same types of numbers began to clamor for recognition in physical problems that had nothing to do with linear measure, the establishment of a new theory of number became an absolute necessity. What we today call real numbers and represent by (ultimately infinite) decimals, were still intuitively conceived as corresponding in some fashion to linear magnitudes. But this was only an intuition and not a well-defined concept; neither the structure of the Euclidean line nor the totality of decimals representing lengths of its finite intervals was well defined. And the fact that not all was well in the foundations of the calculus became a veritable scandal of which even the nonscientific intellectuals became aware (the needling of mathematicians by Bishop Berkeley is the classic example; see Struik, 1948a, p. 178). A situation existed in which the extension of the notion of number, along with its characteristics as an assemblage, was needed. Both hereditary and environmental cultural stresses were at work.

The course of mathematical evolution had brought these early analysts to where they must confront two difficult matters: the nature of mathematics and especially the nature of the infinite in mathematics. A "crisis" existed not unlike that faced by the Greeks when they found that some magnitudes were incommensurable and, in addition, that they were unable to give a satisfactory resolution of the paradoxes of Zeno. Undoubtedly, so far as the nature of number was concerned, the cultural remains of the mystical ideas of the Pythagoreans, as well as of various astrological cults, survived and helped to endow number with an absolute character that it did not have. The necessity for a usable and explicit *definition* of the number concept, corresponding to what Eudoxus did for magnitudes, was not yet realized. Consequently, one was bid to "Seek and ye shall find," rather than to "create," and one sought to discover the nature of a concept whose elusiveness was only a sign of its nonexistence. And it was not until the nineteenth century that it was at last forced on the mathematical world that one must first *define* that totality of numbers which was needed for the foundations of analysis; in short, to give a more precise formulation of what was then only an *intuitively* conceived concept.

Lurking in the background was the necessity for giving a precise formulation of the infinite. For once the concept of real number was available, the *totality* of real numbers must be admitted as a foundation of the calculus. And, as we shall see, this totality proved to have an infinite character of a higher order than that of the totality of natural numbers. One might say that heretofore the mathematician had only encountered the finite, but that now, at long last, he was facing the tremendous leap into the infinite.

To clarify these notions, a method will be sketched by which one can start with the finite decimal fractions and proceed to the concept of the totality of real numbers. For simplicity of exposition, we shall take a "naïve" approach, assuming, in a sense to be made clear later, that every finite decimal is the symbol of a "number." Moreover, we shall temporarily violate the rule that one should distinguish between symbol and the thing symbolized. An extra dividend from this approach to the concept of real number will be the exemplification provided of a phenomenon wherein one starts with a symbolic device (here the finite decimal) and extends it to a broader range of symbolism whose conceptualization is made only *subsequently*. This process occurred repeatedly, of course, in the evolution of number; conceptualizations of 0, $\sqrt{-1}$, and the like, all came later than their introduction as symbolic devices.

1 THE REAL NUMBERS   *Symbolically,* one might think that with the introduction of the decimal point and extension of the place value system to finite decimal fractions, the evolution of the number system that is used in counting and measuring was complete. But consider the fraction $\frac{1}{3}$; does it correspond to a decimal fraction? As anyone who can perform a simple division knows, when 1 is divided by 3, a precise symbolization of $\frac{1}{3}$ does not result, but a series of *approximations:*

$$0.3, \quad 0.33, \quad 0.333, \quad 0.3333, \quad \ldots .$$

One would, therefore, conclude that the finite decimal fractions generally yield only approximations to fractions.

But this difficulty can be avoided and an exact symbolization attained by noticing that such approximations are always *repeating decimals.* The fraction ⅓ yields a decimal fraction in which the repetitions consist of an unending sequence of 3's. A more instructive example is ⁵⁄₇, which yields a decimal fraction in which the digits 7, 1, 4, 2, 8, and 5 are repeated *endlessly* in the order named:

(1)   0.714285   714285   714285   714285 . . . .

Now, a simple symbol may be introduced to indicate this endless array of digits, by placing dots over the digits that repeat; thus

(2)   0.7̇1̇4̇2̇8̇5̇

which will then be supposed to represent the endless array composed of repetitions of 714285 as indicated in (1). Both (1) and (2) are, then, symbols for ⁵⁄₇, except that (2) is precise once the symbol is adopted, whereas in (1) the dots only symbolize that further digits must be supplied, but do not guarantee repetition. (For example, $\sqrt{2}$ is also indicated by 1.414 . . . , but here the dots do not indicate repetition.) Similarly, for ⅓, the symbol 0.3̇ gives a precise symbolization.

Of course not all the decimal part of a number need repeat. For example, the fraction 3123/1400 yields 2.230̇7̇1̇4̇2̇8̇5̇ in which only the digits 714285 repeat. One may easily check that every fraction can be symbolized in this manner, since the usual division algorithm, whenever division does not "come out even," ultimately yields repetition of digits. It is only necessary to consider a fraction p/q in which p and q represent natural numbers such that $p < q$. For if $p > q$, a few steps in the division of p by q yield an integral quotient together with a remainder that is less than q; thus 123/5 gives 24 and a remainder of 3, so that we write 123/5 = 24 3/5. Hence one need only consider in this case the decimal fraction for ⅗, that is, .6, giving 123/5 = 23.6.

Assuming, then, that $p < q$ and that the division does not "come out even," it is inevitable that the partial remainders will repeat. The reason for this is that there are at most q choices for partial remainders at each step. Thus in dividing 3 by 11,

$$
\begin{array}{r}
0.027 \\
11\,\overline{\smash{)}\,3.00} \\
\underline{22} \\
80 \\
\underline{77} \\
3
\end{array}
$$

we get partial remainders 8 and 3, and since we started with the dividend 3, repetition commences. The only possible partial remainders, when dividing by 11 (and assuming that division does not come out even), are the integers from 1 to 10, so that in at most 11 steps repetition will necessarily commence.

Now conversely, every repeating decimal can be represented by a fraction p/q, where p and q are integers. For instance, consider $0.\dot{3}$ ($= 0.3333\ldots$). We know that this results from $\frac{1}{3}$, but supposing that we did not know this, how could we find it out? How do we find the specific fraction it represents? Let us set up the following diagram:

Let $N = 0.\dot{3}$ ($= 0.333\ldots$)   Line 1
Then $\underline{10N = 3.\dot{3}}$ ($= 3.333\ldots$)   Line 2
$10N - N = 3$ (by substracting line 1 from line 2)
Or $9N = 3$, which is equivalent to $N = \frac{1}{3}$

Any repeating decimal can be handled in an analogous manner (although the multiplier will not always be 10), giving a fraction p/q such that, dividing p by q (using the division algorithm), the original repeating decimal with which one started results.

The "discerning reader" may take exception to the last sentence. Consider, for example, the symbol $3.23\dot{9}$. Using the method just described (with 100 as multiplier), the following diagram results:

$$\text{Let } N = \quad 3.23\dot{9} \quad (= \quad 3.23999\ldots) \qquad \text{Line 1}$$

$$\text{Then } 100N = 323.\dot{9} \quad (= 323.99999\ldots) \qquad \text{Line 2}$$

$100N - N = 320.76$ (by subtracting line 1 from line 2, using the symbols in the parenthesis for the substraction)

Or $99N = 320.76$, which is equivalent to $N = 320.76/99 = 32076/9900$ after multiplying numerator and denominator by 100.

But upon dividing 32076 by 9900, one gets 3.24, not 3.23$\dot{9}$! What is "wrong" here? Actually, nothing. For if one proceeds similarly with a decimal ending in a $\dot{9}$, *always* a finite decimal results. And as a result, one is forced to consider the two symbols as *representing the same number*. More precisely, every *finite* decimal may be symbolized by a symbol ending in $\dot{9}$; all one need do is change the last digit to the next smaller digit and follow by $\dot{9}$; then 3.24 becomes 3.23$\dot{9}$ (as we saw above), and 46.271 becomes 46.270$\dot{9}$.[1] For the integers themselves, the same rule applies: 3 becomes 2.$\dot{9}$, and 1 becomes 0.$\dot{9}$ (the reader, if not already familiar with these matters, should check these statements, using the above type of diagrams). Thus every repeating decimal gives either a whole number or a fraction—and, as shown above, every whole number or fraction gives a repeating decimal.

Thus fractions, where now we use "fraction" to include "whole number" (3 as $\frac{3}{1}$, for example), and repeating decimals represent the same numbers; more precisely, if p and q are integers, then dividing p by q gives a whole number or a decimal that ultimately becomes repeating—and this decimal can be shown (as above) to "give back" the fraction p/q; and any decimal that ultimately be-

---

[1] Since the dot convention applies only to the decimal part of a number, a whole number such as 470 is written 469.$\dot{9}$.

comes repeating will (as above) yield a fraction p/q such that if one divides p by q (using the usual division algorithm), one obtains the original decimal with which he started. As stated in Chapter 2, Section 3.1a, the technical name for such numbers is "rational numbers"; any number that may be expressed as the quotient of two integers, that is, in the form p/q where p and q are integers, is called a *rational number*. In view of the discussion above, an alternative definition would be, "the number zero or any number that can be expressed as a decimal fraction that ultimately repeats." An important feature of these numbers is that we can "view," conceptually speaking, their complete decimal symbolizations "to infinity."

1.1 *The irrational numbers and infinity*   In the above manner, there is singled out a whole category of the "measuring," or "real numbers," namely the rational numbers. Are these all? That is, is every real number a rational number? Consider $\sqrt{2}$. If we use the algorithm for taking the square root, we get a decimal representation $1.414\ldots$ in which endless repetition cannot occur, no matter how long the algorithm is used. For as shown above, repeating decimals come only from rational numbers, and it is easy to show that $\sqrt{2}$ is not a rational number.[2] This means that in the case of $\sqrt{2}$, no complete decimal symbolization can be assumed as its representative, unless the conception of an *infinite* decimal, nonrepeating, is accepted. Since $\sqrt{2}$ is not unique in this respect—there are infinitely many nonrational, or *irrational* (as they are technically called), numbers—nonacceptance of such a concept would mean that we must either get along, so far as decimal representation is concerned, with decimal approximations

---

[2] If $\sqrt{2}$ were a rational number, it would be represented by a fraction p/q in which p and q have no common integral factor. Then p^2/q^2 must be 2. But this implies that $p^2 = 2q^2$, and p^2 is an even number. However, if p^2 is even, then so is p, and p^2 must have 4 as a divisor—which implies that $2q^2$ has 4 as a divisor (since $p^2 = 2q^2$). Then q is even. But then both p and q are even, contradicting the fact that p and q have no common integral factor.

or give up the idea of representing such numbers in decimal form.

This is the first time that we have encountered the "problem of the infinite" in discussing the evolution of number. An eminent mathematician, the late Hermann Weyl, termed mathematics "the science of the infinite" (1949, p. 66). We are now on the verge of seeing why. Let us go back to the primitive counting process for a moment. We recall that the earliest counting consisted of an equivalent of one of the systems: "one, many"; "one, two, many"; "one, two, three, many;" and so on. With the development of an adequate symbolism, such as the Babylonian place value system, "many" or its equivalent became unnecessary, since *no matter how large a natural number may be, the place value system is capable of assigning it a unique symbol.* But did the Babylonians, or for that matter the Hindus, Arabs, or other possessors of the decimal place value system, follow up with the conception of an *infinite* totality of numbers? Even with the evolution of a number concept—"number" as noun—did there emerge any notion of numbers so conceived as forming an infinite totality?

Evidently most of the predecessors of the late nineteenth-century mathematicians either had no conception of the natural numbers as forming an infinite class or rejected it (cultural resistance). A notable exception was Galileo (1564-1642), who, in a work published in 1638,[3] not only seems to have spoken of the natural numbers as forming an infinite collection, but resurrected the primitive idea of one-to-one correspondence that we encountered earlier in connection with tallying. For instance, he observed that the squares of numbers are "equal" in number to the numbers themselves; that is, to each natural number n corresponds n^2, so that "counting" the series of squares

$$1, 4, 9, 16, \ldots$$

[3] *Discorsi e dimostrazioni matematiche intorno a due nuove scienze,* Leida, 1638, pp. 32-37 (reference obtained from Bell, 1945, p. 600).

uses up *all* the natural numbers. Now, this makes no sense unless the numbers are conceived as forming a *complete* infinite totality; for otherwise one runs out of squares to assign to natural numbers, inasmuch as the squares grow in size much faster than the numbers themselves. As will be seen later, the same ideas formed the basis of Cantor's successful innovation of "transfinite" numbers; but this was to be 250 years later. E. T. Bell observes (1945, p. 272), "it seems strange that so plain an indication as Galileo's of a feasible attack on . . . the infinite was not pursued sooner than it was. But there is the earlier parallel of the Greek indifference to Babylonian algebra to suggest that mathematics does not always follow the straightest road to its future."[4] A more accurate description of the situation, however, would be as one in which the hereditary stresses were not yet sufficiently great to force such a study; nor would they be until the theory of real functions and related questions *compelled* investigation of the various types of infinite sets—and this was not to occur until the nineteenth century. The great Leibniz (1646-1716) who followed Galileo so closely in history, although also aware of a similar type of correspondence (the correspondence between the natural numbers and their doubles), concluded therefrom that "the number of all natural numbers implies a contradiction."[5] And even the equally great Gauss maintained that "the infinite is merely a way of speaking" (Bell, 1937, p. 556).

This brings up the question of *mathematical existence*. What concepts are permissible in mathematics? Is any limitation to be put on them? Because of the origin of numbers, as well as of geometry, in the world of physical reality, both philosophers and mathematicians have repeatedly sought to justify the "reality" of

---

[4] From *The Development of Mathematics,* by E. T. Bell. Copyright 1945 by E. T. Bell. Used by permission of McGraw-Hill Book Company.

[5] Quote from Bell, 1945, p. 273. The reference given by Bell here is to *Philosophische Werke* (edited by Gerhardt), Vol. 1, p. 338.

mathematical concepts by appeal to physical reality. Thus one may argue that a "huge" number such as

$$10^{10^{10^{10}}}$$

has no "real" existence unless there exists in the physical universe a collection having this number of elements. Similarly, thousands of pages have been devoted to discussions of the question whether Euclidean geometry is "true" or not; in particular, is the "time continuum" represented faithfully by the Euclidean straight line?

Is the validity of a mathematical concept to be judged by its possible connection with some physical reality? To grant this would set up a criterion impossible to apply—too many cases would arise in which the decision regarding validity could not be reached. For example, is π an admissible number? True, it represents the ratio of the circumference of a circle to its diameter, but where in nature can one find a circle? No actual physical "circle" is a circle in the mathematical sense. Moreover, such "numbers" as $\sqrt{-1}$, long rejected by the mathematical world, but ultimately admitted because of the pressure of hereditary stress, eventually became indispensable in the analytic methods of modern sciences such as physics. But the final word, as usual, regarding such questions of "existence" is given by the needs of mathematics; if the need of a concept proves strong enough, the concept will be admitted to mathematical validity. Hereditary stress forces its admittance.

In particular, the invention of the concept of an infinite totality was a result of cultural stress of a hereditary character.[6] Purely

---

[6] It needs to be repeated here that although there is a direct line from those concepts that owed their inceptions to environmental stress, to concepts such as the infinite, the immediate stresses in the latter case were chiefly hereditary. For example, Fourier's studies in the theory of heat led to trigonometric series, which in turn were instrumental in initiating Cantor's study of infinite collections.

mathematical concepts (especially those relating to the analysis that followed the work of Fermat, Descartes, Newton, Leibniz, Cauchy, and others on analytic geometry and the calculus) *forced* a decision on such matters as we are now discussing, especially regarding the notion of *real number*. A "finite" mathematics based on a concept of natural number that envisages only certain initial numbers 1, 2, 3 and so on, together with a rule (adding 1) enabling one to generate as large numbers as one wishes, together with a concept of decimal fraction that allows as close an approximation to a given irrational number as one wishes, may be sufficient in a culture that has advanced only to a stage where such numbers are adequate for scientific purposes. But such theories as are embodied in the calculus and, generally, in real analysis, which were themselves largely a product of environmental stresses exerted by mechanics, physics, and the like, ultimately created hereditary stresses that demanded an "infinite" mathematics for their further development.

Whether infinite totalities "exist" in the physical world has nothing to do with the case. What matters is, do the concepts lead to fruitful mathematical developments? And the answer is that they do; hence their invention. The work of Leibniz and Newton on the calculus led inevitably to questions concerning "infinitesimals," "little zeros," and other vague notions that could only be settled by introduction of a concept of the complete totality of real numbers that forms the basis of the calculus and of all real analysis based thereon. So long as these ideas remained vague, they were legitimate objects for philosophical criticism. But, and more important for mathematics itself, vagueness could (and did) lead to absurdity. It was the hereditary stress felt *within* mathematics, more than external stress of a philosophical or physical nature, that forced a decision regarding the "foundations" of the real number system. So long as one had a mathematics that "worked" in the sense that it provided a satisfactory tool for applications in natural science as well as providing an aesthetically satisfying

An example (unending series of reflections) such as might underlie the *intuitive* notion of infinity. (Charles Eames.)

structure allowing of advances by creatively minded mathematicians, it was found best to follow the path of least resistance and leave well enough alone. The advice of D'Alembert, "Go forward and faith will come to you" (Struik, 1948*a*, p. 220), expresses well the prevailing attitude of the time in which it was uttered. One could always ignore the criticisms of the "nonprofessional" (the "noncognoscenti"), attributing them to failure to understand. But when, as ultimately occurred, the mathematical structure itself shows signs of collapse (owing to such circumstances as the appearance of contradictions, failure to provide adequate foundation for further theory construction, and the like), then "crisis" develops and mathematicians are forced to take thought; hereditary stress then becomes compelling. This is exactly what occurred during the nineteenth century. A situation developed that was quite analogous to that which the Greeks faced—only now the solution could not be geometric. The geometric theory of "magnitudes" was either not an acceptable solution for the new crisis or had to be revamped conceptually and symbolically in order to prove applicable.

Following initial work by Cauchy, the efforts of Dedekind, Weierstrass, Cantor, and others were directed intensely toward providing a rigorous theory of real numbers. Their work involved the assumption of a complete infinite totality of real numbers and employed various approaches, such as "Dedekind cut" classes of rational numbers, equivalence classes of certain sequences of rational numbers ("Cauchy sequences"), and the like. However, these matters are quite technical and are not necessary for our discussion. Instead, we shall give a sketch of the real number system that forms a natural development from what has already been stated concerning decimal numbers.

1.2 *The infinite decimal symbol for a real number*   Let us admit the concept of an infinite decimal and consider an arbitrary infinite decimal of the form

(1) $$a_1 a_2 \ldots a_k . d_1 d_2 \ldots d_n \ldots$$

where $a_1 a_2 \ldots a_k$ is the "integral" part of the number and $.d_1 d_2$
$\ldots d_n \ldots$ the "decimal part" of the number. For example, the
number $247 \frac{1}{3}$ has $a_1 = 2$, $a_2 = 4$, and $a_3 = 7$ (since here $k = 3$),
while d_1, d_2, and in general every d_n, is 3. This number being
rational, we can tell, for every natural number n, exactly what d_n
is. A similar statement holds, of course, for every rational number,
since ultimately its d_n's repeat in groups. However, for irrationals
this is not the case, since for a number like $\sqrt{2}$ we do not know
what $d_{1,000,000}$, say, is. True, if we wanted this digit badly enough,
we could set a computer to work to determine it. But all we can
do is get the finite approximations; admission of the *concept* of an
infinite decimal symbolizing $\sqrt{2}$ does not at all entail our being
able to *view,* in *written form,* all the digits in this symbolization—
such would be out of the question, of course. Another way of
putting this is to acknowledge that so far as our actual symbolic
constructions are concerned, we remain virtually in the same position as the "preinfinite" mathematicians. But this does not prevent
our entertaining the concept of the infinite decimal in the case of
the irrational, if it leads to a fruitful mathematical theory.

Before proceeding, however, attention should again be called
to the ambiguity that we observed during our discussion of the
manner of obtaining fractions from repeating decimals. Consider
the number $\frac{1}{2}$. This may be expressed in decimal form either as

$$(2) \qquad\qquad 0.50000 \ldots$$

or as

$$(3) \qquad\qquad 0.4\dot{9}$$

There is nothing novel in representing a number by more than one
symbol (we already commonly do so, as in the case of the symbols
$\frac{1}{2}$ and 0.5). But here we have two decimal forms for the same
number. Moreover, every "finite" decimal[7] except zero has am-

---

[7] A decimal is "finite" if, in its complete (infinite) decimal representation, all digits d_n are 0 from some point on—that is, for all n larger than

biguous decimal representation. In theoretical work when using the decimal form for numbers, the mathematician usually avoids this ambiguity by conceiving of *all* real numbers (except zero) in their "infinite" form (i.e., 1 as $0.\dot9$, for instance).

But in what sense is a decimal of the form (1) to be conceived as representing a *number?* Disregarding the special case of the natural numbers 1, 2, 3, . . . , which will be discussed more fully later, and concentrating on the "fractional," what is meant by 0.5? To say that it means $\frac{1}{2}$ is only to beg the question. If pushed, one might answer that $\frac{1}{2}$ means "half of something," in which case he has reverted to the notion of the fraction as a "measuring number."

1.3 *The real number as "magnitude"*   There is nothing wrong about conceiving of fractions as representing "measuring numbers," if in so doing one has an intuitive concept for the latter. In doing so, one returns essentially to the Greek idea of "magnitude," of course. This was a very useful conception, not just for the Greeks, but for the early analysts who, after the introduction of analytic geometry, thought of numbers as measured off on a line. And it will be advisable for the reader who has not previously encountered the infinite decimal to make use of this conception before attempting to understand the modern point of view.[8] So let us show how one may give an interpretation of (1) as a measuring number. For a number like $\sqrt{2}$ we can go back to its origin as the measure of the diagonal of a unit square. But what about a general infinite decimal like (1) that has so such ready-made connotation?

It will be presumed that one knows how to measure off a length equal to a given (natural) number of units, such as 5 units,

---

some fixed natural number. Customarily the 0's are not written down; for example, 2.645, representing 2 and 645/1000.

[8] According to N. Bourbaki, 1960, p. 160, during the sixteenth century R. Bombelli actually attained to a geometric definition of the *complete arithmetic* of real numbers by reverting to the Greek idea of representing numbers by line lengths.

using any familiar unit of measure. Then we may concentrate on interpreting a "pure" decimal:

(4) $0.d_1d_2d_3 \ldots d_n \ldots$

For to get a line of length (1), $a_1a_2 \ldots a_k$ units can first be measured off and then extended by the "length" that will be defined for (4). It must be emphasized, however, that the "measuring off" is purely a conceptual act, since no actual measuring instruments exist that will measure off even a given number of units perfectly.

Let us consider a line segment S of unit length and label one end of it with a 0 and the other with a 1; we can safely use the terms "left" and "right" if we call the end labeled 0 the left end and the other the right end. Divide S into ten equal parts and label the points of division 0.1, 0.2, . . . , 0.9 successively:

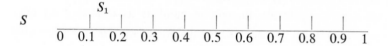

Let us consider a particular case, say the decimal part of π:

(5) $0.14159 \ldots$

Guided by the fact that the first digit in (5) is 0.1, we begin by selecting the interval from 0.1 to 0.2 on S and label it S_1.

We next divide the interval S_1 into ten equal parts, labeling

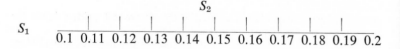

the points of division 0.11, 0.12, . . . , 0.19 successively. Guided by the fact that the first two digits of (5) are 0.14, we select the interval from 0.14 to 0.15 and label it S_2.

At the next stage of this process we divide S_2 into ten parts, labeling the points of division 0.141, 0.142, . . . , 0.149; and since the first three digits of (5) are 0.141, we select the interval from 0.141 to 0.142 and label it S_3.

The theoretical continuation of the procedure should be clear, and anyone familiar with the process of definition by mathematical induction will see how to define a complete infinite series $S, S_1, S_2, . . . , S_n, . . .$ of line segments, each containing its successor. It should be noted, moreover, that the number (5) determines this series of intervals uniquely; any different decimal will give a different series of intervals.

Now it is a fundamental property of the Euclidean line that such a sequence of intervals has *exactly one* point P, that is common to them all. And if the point labeled 0 is called A, then the length (measure) of the interval AP may be considered to be the number represented by (5). A better way of looking at this is to consider that the decimal (5) *constitutes a unique label* (i.e., *symbol*) *for what the Greeks would have called the "magnitude"* AP. For it is easy to reverse the above process: Given the point P, the successive digits of (5) are determined by dividing S into ten equal parts and locating the interval S_1 containing P; then dividing S_1 into ten equal parts and determining the interval S_2 containing P —at the same time "reading off" the left-hand labels 0.1 of S_1, 0.14 of S_2, and so on, in order to determine (5). Again it must be emphasized that we do not carry through any such construction *physically*, either in defining AP in terms of (5) or the converse thereof; we only *define* the corresponding concepts.

Since the general process should be clear from the way the special case (5) has been handled, there is no need to repeat it for the general case (1). So assuming that it is known how, in the above manner, a "magnitude" is defined for any decimal of form (1), it can then be concluded that there exists a (1-1)-correspondence

between the real numbers and the points of a Euclidean line. For
if L is such a line, a specified point of L may be labeled 0. Assuming the convention made regarding what "right" and "left" directions of L mean,

$$L \quad \overline{\begin{array}{ccccccc} -3 & -2 & -1 & 0 & 1 & 2 & 3 \end{array}}$$

a unit may be selected and points to right of 0 that are 1 unit, 2
units, and so on, from 0, labeled 1, 2, 3, . . . (see figure). Points
to the left are labeled similarly with negatives of the natural numbers. Consider any number $a_1a_2 \ldots a_k .d_1d_2 \ldots d_n \ldots$ of form
(1). If it is positive, the unit interval immediately to the right of
the point labeled $a_1a_2 \ldots a_k$ can be labeled S. In S the point P
corresponding to the decimal $d_1d_2 \ldots d_n \ldots$ is then defined as
indicated in the process above. (For a negative number, the point
P would be defined to the left, of course, and the labels in the
figure for S above would run from right to left instead of from left
to right.) Since the reverse of the process has been described, the
(1-1)-correspondence is established. It is this correspondence that
is at the basis of analytic geometry. (The assertion of the correspondence is sometimes called the Cantor axiom.)

Observe that it was precisely this concept of providing a
unique numerical symbol for a magnitude that the Greeks lacked.
With a suitably cipherized sexagesimal system, which could have
been devised by combining the Babylonian sexagesimal place value
system with, for example, Ionian numerals, the Greeks might have
provided themselves with a symbolic tool of far greater utility than
the geometric constructions to which magnitudes were limited.

1.4   *The real numbers based on the natural numbers*
The dependence that mathematicians, of the centuries preceding
the nineteenth, placed on geometry and geometric intuition has
already been remarked on in Chapter 2, Section 5.3. But not
only did geometric intuition prove to be an unreliable guide;[9] it

[9] This was not the fault of the geometry, to be sure; it was only that the
rigorous analysis of the line from the standpoint of what is now called
point set theory had not been made.

seemed to offer no solution to fundamental problems concerning differentials, continuity of functions, and the like in which computations are involved. This is not to say that, properly handled, geometry could not have provided a solution;[10] only that the direction that mathematical analysis was taking led clearly to a demand for "arithmetizing" the real number system. And it was this that the nineteenth-century analysts such as Weierstrass, Dedekind, and Cantor accomplished. By "arithmetizing" the real number system is meant basing the concept of the real number system on the natural numbers and their arithmetic (of addition, multiplication, etc.), making no appeal to geometry.

It is not difficult to start with the arithmetic of natural numbers and give definitions, based thereon, of the rational numbers and their arithmetic; the details will not be given here, except to remark that it may be done using ordered pairs (p, q) where p and q are natural numbers. (See, for instance, Wilder, 1965, Chapter VI, Section 3.2.) Then, assuming that one has defined the rational numbers and their arithmetic in this way, one can proceed to define "real numbers" in several ways. For example, limiting the discussion again to the "pure" decimal, a real number can be defined as a sequence of rational numbers of the following type:

$$(6) \qquad .d_1, .d_1d_2, .d_1d_2d_3, \ldots, .d_1d_2 \ldots d_n, \ldots$$

in which each term differs from its successor only in the fact that in its decimal representation a new digit has been added to obtain the successor. This would correspond in an obvious manner to the decimal part of a real number as symbolized in (1). But, one may ask, what advantage has (6) over (1)? The answer is that the elements that constitute (6) are all familiar rational numbers, while (1) is, a priori, meaningless. Of course, to get a complete theory (including operations of addition, multiplication, etc.), one must still define addition, multiplication, and so on, of such num-

[10] Recall the note concerning R. Bombelli (footnote 8). Such a solution would no doubt not have been symbolically satisfactory, however.

bers as (6) (as well as establish what is meant by one real number being "less than" another—a concept that is rather obvious when one uses the previous interpretation in terms of "magnitude").

One may still prefer to conceptualize real numbers as measuring numbers (magnitudes), based on the (1-1)-correspondence with points of the Euclidean line as outlined in Section 1.3. For present purposes this will be sufficient, perhaps even preferable. But this would not be the case if one were going on to define the arithmetic of the real numbers; this can be done more satisfactorily in terms of the rationals. Indeed, what we have just pointed out, that is, the feasibility of defining real numbers in terms of rationals, hence in terms of the natural numbers, is of great importance to mathematics. It enables one to free analysis, which is ultimately based on the real number system, from geometric intuition.

Despite this, however, the working mathematician actually uses *both* concepts—that of the real number as magnitude and the real number arithmetized [as a sequence of type (6), for instance]. He has found, through a study of the collections of points on a Euclidean line, both how to conceptualize satisfactorily many analytic notions (e.g., integration) and how to avoid the errors to which the unanalyzed Euclidean line can lead.[11] This has led to a separation of the real number concept relative to (1) the structure of the real line (i.e., its topology) and (2) its arithmetic and algebraic properties. In the former case, (1), the real number is essentially treated as a geometric entity; in the latter, (2), as the arithmetized entity based on the natural (via the rational) numbers. Nevertheless, through the device of labeling points on a line with arithmetized real numbers ("Cantor axiom"), the concepts are usefully combined by the working mathematician. The term "linear continuum" is ordinarily used to denote the Euclidean line, and "real number continuum" to denote the real numbers (disregarding operations); as our discussion shows, they are conceptually equivalent.

[11] From such a study arose not only the modern theory of sets of points, but the founding of modern set-theoretic topology.

2   THE CLASS OF REAL NUMBERS   The argument given in Section 1.1 for overcoming the cultural resistance to the concept of an infinite class such as that of the natural numbers, or of the infinite decimal, can be profitably recalled at this point. For the demands of nineteenth-century analysis required not just the concept of an individual real number, but of infinite classes, or *sets* (to use the usual synonym), of real numbers. In particular, the concept of the set of *all* real numbers—or, equivalently, the set of all points on the Euclidean line—constitutes an infinite set that may be termed to be of a higher-order of abstraction than that of the set of all natural numbers. Indeed, if one considers the concept of a complete infinite totality like that of the natural numbers as rather awesome, he may prepare to magnify his fund of awe to "higher dimensions." For it turns out that there are so many of these real numbers that they could not possibly be "labeled" with natural numbers, even if the complete totality of natural numbers were used!

In order to make clear just what is meant by the last sentence, let us consider again the elementary counting process. In counting a collection of, say, six objects, the person counting would ordinarily point (literally or subjectively) to each object in succession and pronounce an appropriate number word (natural number). In terms of our English number words, he would point to one object and say "one," then to another object and say "two," and so on until he had pointed at every object and said, finally, "six." In so doing he has "labeled" each object with a (natural) number word, being careful (1) not to label an object more than once and (2) to label every object. In technical mathematical terms, he sets up a (1-1)-correspondence between the objects and the natural numbers from "one" to "six."

Now when one has accepted the concept of the set of *all* the natural numbers—that is, the infinite totality—it is legitimate to ask about any given infinite collection (like that of the real numbers) whether its objects—"elements" is the technical word—can be labeled with symbols of the natural numbers, that is, with the numerals 1, 2, 3, and so on. Of course it would not be asked whether this can be done *physically* or by consecutive mental acts ("pointing"), as can be done in the case of a collection of six objects; this would be manifestly impossible. So it is better to ask only whether the (1-1)-correspondence can *exist*. Here by "exist" is meant to exist in any sense whatsover—possibly by giving a law that automatically sets up the correspondence, as Galileo did for the squares of natural numbers (by the law making n^2 correspond to n for each natural number n). Or, again, would it be legitimate to base a mathematical theory on the assumption that such a correspondence may be set up? There are infinitely many natural numbers, so perhaps the answer is positive.

Suppose that such a correspondence between the natural numbers and the real numbers does exist. A convenient way of symbolizing this would be to say that to each real number r has been assigned a "label," namely a numeral $n,$ so that the real number that is labeled with a numeral n can be denoted by r_n. In effect, then, there exists a sequence

$$r_1, r_2, \ldots, r_n, \ldots$$

of symbols representing the real numbers; every real number is represented by some r_n in this sequence.

In what follows, a unique decimal representation is needed for each real number, so the convention will be to represent each real number by its "infinite" decimal form e.g., for $\frac{1}{2}$ we use $0.\dot{9}$ instead of 0.5). (Zero is represented by $0.000 \ldots$). Only the decimal fraction portion of each such symbol need be considered, so let the decimal part of a number r_n be denoted by

$$.d_1{}^n d_2{}^n \ldots d_n{}^n \ldots$$

where the nth digit of r_n is symbolized by $d_n{}^n$; the index n in $d_1{}^n$, $d_2{}^n,$ and so on, serves to indicate that a digit of the number r_n is being referred to. We can conveniently think of all the decimals obtained from the real numbers in this way as assembled in an array as follows:

(Decimal part of r_1): $.d_1^1 d_2^1 \ldots d_n^1 \ldots$

(Decimal part of r_2): $.d_1^2 d_2^2 \ldots d_n^2 \ldots$

(7) . . . .

(Decimal part of r_n): $.d_1^n d_2^n \ldots d_n^n \ldots$

. . . .

Now, each d_n^n is a label for one of the digits 0, 1, . . . , 9. For each natural number n, let a digit d_n be defined as follows: If d_n^n is 1, let d_n be 2; if d_n^n is not 1, let d_n be 1. (Thus, if r_1 is $0.49\dot{9}$, d_1 will be 1, since d_1^1 is 4, hence not 1; if r_2 is the decimal part of $\sqrt{2}$, that is, $0.414 \ldots$, then d_2 is 2, since d_2^2 in this case is 1.) This "law" defines a decimal

(8) $\qquad\qquad .d_1 d_2 \ldots d_n \ldots$

hence the symbol for a real number r. But consider: Since r is a symbol for a real number, it must be in the array (7) corresponding to, say, r_k. But then the kth digit in (8) is different (by the above law defining r) from d_k^k, the kth digit of r_k, so that r and r_k cannot symbolize the same number. To suppose, then, that the real numbers can be labeled with natural number numerals, hence set up in an array like (7), leads to contradiction.

In view of the preceding paragraph, it can only be concluded that the real numbers cannot possibly be "counted"—that is, labeled—with natural number numerals. (For to assume that they could be so labeled yields a way of defining a real number r which is *not* labeled, as shown above; hence a contradiction.)

2.1 *The Cantor diagonal method*   The *method* used to define the number r symbolized in (8) is usually called the Cantor diagonal method, because it was used for this purpose by the late

Georg Cantor, 1845-1918, "founder" of the mathematical theory of the infinite. (Georg Cantor, *Gesammelte Abhandlungen,* Verlag von Julius Springer, Berlin, 1932.)

German mathematician Georg Cantor (1845-1918) and because it utilizes the elements (digits) $d_n{}^n$ on the diagonal of the array (7). (Incidentally, as he framed the argument the ambiguity that we have avoided by using a unique decimal representation was not recognized; for not to do so might conceivably allow r in the above proof to represent the same number as r_k in that r might be a "finite" decimal form of r_k).

It should be noted that in defining the number r, the *only* digit of the first row of the array (7) that was used was d_1^1; similarly, the *only* digit of the second row that was used was d_2^2; and so on down the diagonal of (7) proceeding "southeast" from the upper left-hand corner. That is, of each infinite decimal only one of the digits needs to be known in order to define r. Consequently, there has evolved from the method a more general method (not depending on an array of type (7)), which is called simply a "diagonal method," that is employed whenever one wishes to define a new entity, E, in terms of a collection of classes C_i and where the definition of E makes use of precisely *one* element from each class C_i. In the above, the entity E was the number r, and the classes C_i were the rows of digits in (7). This "diagonal method" is not a new *logical* principle; it is a purely *ad hoc* procedure, set up for the specific purpose of defining something—the number r in the case of the Cantor diagonal method—and subsequently found useful in other situations of a more general type. This kind of phenomenon occurs again and again in mathematics; in order to solve some special problem a novel method is introduced which is subsequently found to generalize to other situations and a new method of proof.

For an elementary example of the use of a diagonal method, suppose that one is asked to select sums of money from a collection of coins of four different denominations. Specifically, suppose that the coins are United States currency, the four denominations being "penny," "dime," "quarter," and "dollar." One selection would be "dime," another would be "penny, dollar," and another would be "penny, dime, dollar." Each of these is a different selection; moreover, another selection would be the empty one—that in which none of the denominations is chosen. Now, no special proof method is needed to show that more than four selections are possible; four selections have already been indicated, and more are clearly possible. But suppose that one is asked to give a method whereby *no matter how four selections have already been made, a new one different from each of them can be made.*

It may be done as follows. Let C_1, C_2, C_3, C_4, denote any four selections. In order to make a selection that is different from each of these, we proceed in the following way: Consider the denominations one at a time; first, "penny." Is "penny" in C_1? If it is, then do *not* select it for C, but if it is not, then select it for C. Next, consider "dime"; if it is in C_2, do not select it for C, but if it is not, then select it for C. And so on. Then C is different from each of the given selections C_1, C_2, C_3, C_4, since it differs from each *in at least one denomination*. Notice the analogy to the way in which the number r was selected above; r was selected so as to differ from each of the numbers in the array (7) in at least one digit (the digit $d_n{}^n$ on the diagonal).

The process just described constitutes a diagonal method, and it can be employed similarly for any natural number n; more precisely, given a collection S of n things, if n selections are made from S, a further choice can be made that is different from all those already made. That is, *from a collection S of n things, more than n different selections can be made*. This is a theorem that holds for every natural number n, and its proof may be given just as that for the case of four objects above.

3 TRANSFINITE NUMBERS; CARDINAL NUMBERS   The important feature of the italicized statement just made is that it does not depend on any particular value of n; it holds for *every possible* value of n. For a particular value of n, it would be a somewhat trivial statement; its *generality* makes it important. Moreover, a similar statement can be made for collections that are *not finite*. But in order for such a statement to make sense, the terms used must be made precise, particularly the words "more than" (as well as a meaning for n).

Harking back to the quotation from Galileo, observe that he said "the squares of numbers are equal to themselves." By this he evidently meant that the correspondences of 1 to 1, 2 to 4, 3 to 9, and in general n to n^2 (each natural number corresponding to its square) constitute a (1-1)-correspondence between the elements of the set, N, of *all* natural numbers and the elements of the set, S, of *all* their squares. For finite sets such a correspondence between the elements of two sets would mean that the two sets have the same number of elements. But clearly there are "more" natural numbers than there are squares of them, since after deleting the elements of S from N we have left an infinite set of numbers—2, 3, 5, 6, . . .—the "nonsquares."

This must seem very confusing to one who learns it for the first time; perhaps even contradictory (as it did to Leibniz and others). It is no wonder that Leibniz deemed the concept of the set of *all* natural numbers, or of any actually infinite set, to be untenable, if mathematics is to be preserved from contradiction. The difficulty lay in the meanings to be attributed to such terms as "more than" and especially to what one is to understand by "number" of an infinite set.

The disagreement between those who shared Galileo's opinion of the nature of the infinite and those who opposed it did not form a hereditary stress strong enough to force a settlement of the problem until the latter half of the nineteenth century when the need to define more precisely the concept of the linear continuum also focused attention on the nature of the infinite. It was Cantor's great achievement not only to solve the problem but to introduce order into the chaos of the infinite. The most basic accomplishment that he achieved was to show how to extend the concept of (natural) number to infinite collections.

The reader will probably have noticed that not yet, in the course of this survey, has there been given a precise definition of "natural number." As the development of the notion of number has been followed through history, all that has been observed is the evolution of certain concepts regarding "counting numbers," and their extension to "real numbers," or "measuring numbers." These "counting numbers" were created by cultural stress, as were also the "measuring numbers." And the needs of mathematical analysis forced the introduction of a more precise characterization of the measuring numbers, in the form of the "real numbers," but the sketch above of how these may be defined in terms of the natural numbers assumed knowledge of the natural numbers. The "transfinite" numbers defined by Cantor, which will be described below may be considered as a further step in the evolution of the "counting numbers," of which we have, up to now, seen only the natural numbers.

3.1 *Extension of "counting numbers" to the infinite*
First, the convention will be made that if there exists a (1-1)-correspondence between the elements of two collections, they are to "have the same number" of elements. For finite sets, this agrees with the intuitive notion of "counting numbers" which was re-marked upon in Chapter 1, Section 1.2d in regard to the primi-tive use of tally sticks or similar devices. Lacking an adequate set of number words, one can compare two collections by the tallying process even when the two collections are so far removed from one another as to make direct comparison impossible. All that is necessary is to gather a convenient third collection, C, of portable objects (straws, pebbles, or notches on sticks, for instance) whose elements have been put in (1-1)-correspondence with the elements of one of the given collections; then transport it to the location of the other collection and see if it is possible to set up a (1-1)-correspondence between its elements and the elements of the por-table collection C. If it is, one knows that the two original collec-tions have the same number of elements.[12]

Cantor's fundamental idea was that the same criterion—of (1-1)-correspondence—may be used for comparing *infinite* sets.[13] As an instructive example, consider the set of all positive rational

---

[12] The use of the third collection C in this manner involves a principle that mathematicians call the transitive law. A relation, R, between collec-tions, is called transitive if, given three collections A, B, and C and the relation R between A and B as well as between B and C, it follows that the relation R holds between A and C. The reader can probably think of many examples of such relations.

[13] Recall Bell's comment (Section 1.1) regarding Galileo's observations.

numbers and imagine "fractional" symbols for them placed in a square array, on the first line of which are all fractions with numerator 1, on the second line all those of numerator 2, and so on:

$$\begin{array}{ccccc} \dfrac{1}{1} & \dfrac{1}{2} & \dfrac{1}{3} & \dfrac{1}{4} & \dfrac{1}{5} & \cdots \\[2ex] \dfrac{2}{1} & \dfrac{2}{2} & \dfrac{2}{3} & \dfrac{2}{4} & \dfrac{2}{5} & \cdots \\[2ex] \dfrac{3}{1} & \dfrac{3}{2} & \dfrac{3}{3} & \dfrac{3}{4} & \dfrac{3}{5} & \cdots \\[2ex] \dfrac{4}{1} & \dfrac{4}{2} & \dfrac{4}{3} & \dfrac{4}{4} & \dfrac{4}{5} & \cdots \end{array}$$

. . . .

There are many different symbols for the same number when fractions are exhibited in this way, of course; thus these on the main diagonal

$$\frac{1}{1} \quad \frac{2}{2} \quad \frac{3}{3} \quad \frac{4}{4} \quad \cdots$$

are all symbols for the number 1. But this will cause no difficulty and it is very convenient to use this kind of array, since in each "cross-diagonal," for instance, in the third cross-diagonal where lie

$$\frac{3}{1} \quad \frac{2}{2} \quad \frac{1}{3} \; ,$$

the sum of numerator and denominator is constant, being 4 in the case cited. Now, if a path like that indicated in the figure below is followed,

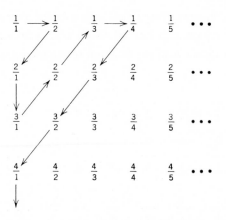

every symbol in the array will be crossed just once. And if a new symbol a_n is assigned to each symbol in the array as the path crosses it (thus a_1 to $\frac{1}{1}$, a_2 to $\frac{1}{2}$, a_3 to $\frac{2}{1}$, a_4 to $\frac{3}{1}$, a_5 to $\frac{2}{2}$, a_6 to $\frac{1}{3}$, and so on), then the sequence

$$a_1, a_2, \ldots a_n, \ldots$$

contains one symbol for each of the symbols in the original array. Moreover, if repetitions of symbols for the same number are deleted—a number once symbolized having all its symbols deleted thereafter—there results a new sequence

$$b_1, b_2, \ldots, b_n, \ldots$$

which contains exactly one symbol that can be assigned for each and every positive rational number.

If now a correspondence is made in which the rational number symbolized by b_n corresponds to the natural number n, there is defined a (1-1)-correspondence between the positive rational numbers and the collection N of natural numbers. [As an exercise, the reader might show that one can go a step further and define a (1-1)-correspondence between *all* the rationals—positive, negative, and zero—and the natural numbers.]

If, then, existence of (1-1)-correspondence is accepted as a criterion for comparison of "size" or "numbers" of sets, two sets having "the same number" of elements if there exists a (1-1)-correspondence between their elements, then the set, F, of all rational numbers has the same number as the set N of natural numbers. This suggests that "number" of a set be conceived as the same concept as the *size* of the set. The latter is only an intuitive notion, of course, but undoubtedly corresponds to the primitive meaning of "number." Whether it is satisfactory for infinite sets, however, as it certainly is for the finite sets that are encountered in physical reality, can only be decided by noticing how it bears out what the intuition demands.[14]

The first step in making such a test is to assign symbols (numerals) for the numbers of known infinite sets. To the set N of all natural numbers, Cantor assigned the "numeral" $\aleph_0$, called aleph-null ($\aleph$ being the initial letter of the Hebrew alphabet). To say, then, that a set S has $\aleph_0$ elements or that the number of elements in S is $\aleph_0$ is to imply that between the elements of S and the natural numbers there exists a (1-1)-correspondence. Notice the analogy to the finite case when one says, for instance, that a set has 6 elements.

Many infinite sets have the number $\aleph_0$, *but not all*. As has already been observed in Section 2, the set, R, of all real numbers cannot have the number $\aleph_0$. For the real numbers, as shown in Section 2, cannot be in (1-1)-correspondence with the natural numbers. Consequently a different symbol must be assigned for the size of R—the one ordinarily used is the letter c (c is the initial letter of the word "continuum," in the phrase "real number continuum").

How do $\aleph_0$ and c compare? Is one "bigger" than the other in the same way that 3 is bigger than 2, for instance? Such a question

[14] Mathematicians have devised more sophisticated ways to define "number," but these are based on systems of axioms for set theory that have to be carefully delineated so as to avoid inconsistencies.

makes no sense unless a meaning for "bigger than" is agreed on. Early mathematicians never were forced to *define* the statement "3 is bigger than 2," for such statements were the product of cultural evolution—what might be called "mathematical artifacts" —rather than of a conscious invention process such as Cantor had to follow.[15] But if the notion of number is to be extended to the "infinite," not only must fundamental terms and relations be given suitable meanings, but it must be done so that they will "work" for the finite cases that have been inherited in the culture. Everyone "knows" what it means to say "3 is bigger than 2," for it is intuitively clear. But as a guide when venturing into the new territory of the infinite, intuition may not be entirely safe. It has already been remarked that the set of all natural numbers is "bigger than" the set of all squares of natural numbers, and similarly F is "bigger than" N, in the sense that if all the natural numbers are deleted from F, there are still infinitely many fractions left. And for finite sets, such a circumstance would surely indicate inequality of size. But having so conceived of "number" as to make N and F have the *same* size, which has been symbolized by $\aleph_0$, whereas R is of different size, a different criterion must be invented for "bigger than"; besides, it is not generally the case that of two sets to be compared, one can be considered as *part* of the other, as is the case with N and F.

An alternative way of comparing two finite sets is the following. Suppose that A and B are two finite sets, and we find that there exists a (1-1)-correspondence between the elements of A and the elements of a part of B, but not conversely; then certainly A has fewer elements than B. The number 3 is certainly bigger than the number 2 in this sense, for if A has 2 elements and B has 3, there is a (1-1)-correspondence between the elements of A and

---

[15] Ask the "man in the street" what "3 is bigger than 2" means and he is likely to reply "because it is"; that is, if he replies at all. More likely he will register a look of disbelief that one should ask such a "stupid" question.

two of the elements of B, but a converse of this situation is impossible (there not being three elements in A to make correspond to the elements of B). Now, Cantor found that this is a criterion that will also "work" for infinite sets; another way of putting this is to say that it is a criterion that is "consistent" with the concept of number proposed above. For instance, applying it to c and $\aleph_0$, it turns out that (as would be expected) c is bigger than $\aleph_0$: for there does exist a (1-1)-correspondence between N and part of R (let the part of R used be either N or F, considered now as sets of real numbers—either will do), but if there were a (1-1)-correspondence between R and part of N, it would easily follow that there was a (1-1)-correspondence between the elements of R and all the natural numbers—which has been shown above to be impossible. It can be seen now why, when decimal symbolization is extended to all the real numbers, the concept of an infinite decimal is unavoidable. For using finite decimals and repeating decimals (which are finitely representable using the dot notation; cf. Section 1), only $\aleph_0$ symbols are obtainable, whereas c symbols are necessary in order to assign a unique decimal to each real number.

If the customary symbol $<$ for "less than" is used, we write

$$\aleph_0 < c$$

just as we write $2 < 3$. And the properties satisfied by this relation are precisely like the traditional "$<$" relations for the natural numbers. For example, the "transitive law" that was mentioned above (if m, n and r are natural numbers such that $m < n$ and $n < r$, then $m < r$) continues to hold for the numbers of infinite sets. Of course, this would not be very meaningful if $\aleph_0$ and c were the *only* "transfinite" numbers. But *there are infinitely many transfinite numbers*. For the proof (in Section 2) that if a finite set S has n elements, then the number of selections that can be made from S is greater than the number of elements in S, extends almost without change to infinite sets. (The mathematician uses the term "subset" instead of "selection"—the latter has too many physical connotations and nobody "selects" the subsets as a rule.)

In particular, it can be proved that the number of possible subsets of the set, N, of natural numbers is exactly c. And the number of possible subsets of the set, R, of all real numbers, is a number sometimes denoted by f, since it is also the number of a certain class of functions defined over R. And so on. Generally speaking, the transfinite numbers of the type we have been describing, along with the natural numbers and zero, are *cardinal numbers* in the sense described in Chapter 1, Section 1.2b. The cardinal numbers may now be divided into the two classes: *finite cardinals,* constituted by the natural numbers and zero, and *transfinite cardinals,* the cardinal numbers of infinite sets. Although it would take us beyond our goals to follow the details, it should be mentioned that the arithmetic of the finite cardinals, such as addition, multiplication, and raising to cardinal powers, has been extended to the transfinite cardinals. And just as there are still many unsolved problems concerning finite cardinals (number theory), there remain many unsolved problems regarding both existence and properties of transfinite cardinals. Only very recently has an answer been found to the question, Does there exist a number between $\aleph_0$ and c, that is, can there be a number λ such that $\aleph_0 < \lambda < c$?[16]

3.2   *Transfinite ordinal numbers*   In Chapter 1, Section

---

[16] Cantor conjectured that the answer to this question would be negative, and many later investigators surmised that this would ultimately be proved on the basis of generally accepted axioms for set theory. Meanwhile, the assumption that no such number λ exists became known as "the continuum hypothesis," and the solution of many mathematical problems was based on the assumption. However, work of Gödel (1941) and of Paul J. Cohen (1963) has finally established that the continuum hypothesis is an independent axiom of set theory; consistent set theories may either include it or deny it, thus putting set theory in a position like that of classical geometry with respect to the parallel axiom—consistent geometries may include it or deny it.

1.2b it was observed that the natural numbers also play an "ordinal" role; that "January 3" means "the third day of January," for example, and seat numbers in a stadium fulfill an ordinal role. But while a statement such as "I shall sit in seat 4 from the aisle" can be equivalently stated in the form "There will be 3 seats between my seat and the aisle," without any misunderstanding due to the fact that the "4" in the first statement is an ordinal number while the "3" in the second is a cardinal number, an analogous use of cardinals cannot be permitted without misunderstanding. For example, the statement that "the number of all natural numbers is $\aleph_0$" does not disclose that they have a certain order; and as was shown in Section 3.1, the number of all rational numbers is also $\aleph_0$. But in their order of magnitude, the rational numbers form an ordered set totally unlike that of the natural numbers when the latter are considered (as they usually are) in their order of magnitude. Thus between 1 and 2 there are no other natural numbers, but between any two rational numbers there are infinitely many other rational numbers.[17] Hence the symbol for a transfinite number, such as $\aleph_0$, *cannot be used as a symbol for an ordinal number* as may be done with the finite counting numbers. On the transfinite level, the distinction between "cardinal number" and "ordinal number" becomes of paramount importance. Cantor recognized this and also that in addition to the cardinal numbers he had to define and study another class of transfinite numbers to be called "ordinal numbers." For the same transfinite *cardinal* number (which, we recall, only measures size) there had to exist infinitely many different corresponding ordinal numbers, which is another way of saying that the same infinite class can be ordered in infinitely many essentially different ways. Thus the natural numbers can be ordered in their usual order of magnitude; but they

---

[17] For instance, between 0 and 1 as rationals, there is ½; between ½ and 0 there is ¼. In brief, between 0 and 1 are, among others, *all* the rational numbers representable in the form $1/n$, where n stands for natural number.

can also be ordered so that (1) each odd number precedes *all* the even numbers while (2) the original order is maintained as between two odd numbers or as between two even numbers. Schematically, one can indicate the ordering described thus:

$$1, 3, \ldots, 2n + 1, \ldots : 2, 4, \ldots, 2n, \ldots$$

4 WHAT IS NUMBER?   It should by now be evident that the question "What is number?" is not meaningful unless the word "number" is qualified in suitable manner. The counting numbers ultimately evolved to the entire class of *cardinals,* finite and transfinite. But as has just been indicated, in Section 3.2, these same numbers—but now taken in their role of *ordinals*—can be extended to a distinct class of transfinite ordinals.

On the other hand, if the natural numbers are viewed as measuring numbers, they extend to the entire class of *real numbers.* And these are not all the types of "number" that have evolved from the ancient Sumerian-Babylonian "number science." A complete history would describe the manner in which negative numbers and complex numbers forced their way into mathematics. The notion of negative number was evidently found useful by the Hindus,[18] whose algebra was influenced by the last of the great Greek mathematicians of antiquity, Diophantus of Alexandria (c. A.D. 300). But it was not until about the seventeenth century that mathematicians began to recognize negative numbers as "valid" numbers. As one would expect, from the way in which the decimal system evolved for instance, there were isolated cases of bold souls who toyed a bit with them. Thus Cardan in his *Ars Magna* divides numbers into *numeri veri,* which were the "real numbers" of his day (including natural numbers, as well as positive fractions and some irrationals), and *numeri ficti,* or *numeri falsi,* which were the negative numbers and the square roots of

---

[18] Some have maintained (e.g., R. C. Archibald) that the Babylonians handled negative numbers.

negative numbers. But he is said to have cautiously admitted the latter on occasions when they either admitted interpretation or could be manipulated so as ultimately to produce "numeri veri." Historians also seem to agree that Albert Girard (1595-1632) not only admitted negative numbers but anticipated Descartes in the use of the symbols $+$ and $-$ to indicate opposite directions on a line.

It would be difficult to separate out the contributions of various cultural forces, such as hereditary and cultural stresses, to the ultimate admission of negative real numbers and expressions involving $\sqrt{-1}$ ("complex numbers") to respectable mathematical standing. On the one hand, a complete theory of algebraic equations demanded their admission; otherwise one cannot state (as does the famous Fundamental Theorem of Algebra) that an nth degree equation has n roots.[19] On the other hand, the demands made of mathematical analysis by the development of the physical sciences were undoubtedly influential in forcing the creation of a complete theory of complex numbers. Ingenuity and boldness in the designing of "natural laws" are not unlikely to foster similar qualities in the invention of the relevant mathematical structures, especially (as was usually the case in those days) when both activities are cultivated by the same person. Accordingly as the need for new number concepts is created by either cultural or hereditary stress (or a combination of both), such concepts will be invented irrespective of objections by "realistic" individuals who may protest their "fictitious" character.

In modern algebra, number systems take a wide variety of forms according as mathematical theory or new applications exhibit a need for them. A "number system," generally (and

---

[19] Struik observes (Struik, 1948a, p. 114): "It is a curious fact that the first introduction of the imaginaries [i.e., numbers involving $\sqrt{-1}$] occurred in the theory of cubic equations, in the case where it was clear that real solutions exist though in an unrecognizable form, and not in the theory of quadratic equations, where our present textbooks introduce them."

roughly) speaking, is any collection whose elements may be combined by two operations denoted by $+$ and $\times$ that satisfy certain elementary properties. These properties are the analogues of those satisfied by the addition $(+)$ and multiplication $(\times)$ of common arithmetics, for example, the arithmetic of the natural numbers or the arithmetic of the real numbers. In particular, if a, b, and c are natural numbers, then $(a + b) + c$ and $a + (b + c)$[20] yield the same number—the so-called associative law of addition. And $a \times (b + c)$ must give the same number as $(a \times b) + (a \times c)$ —the "distributive law." But in the general number systems of modern algebra, the objects added and multiplied may not be what one would ordinarily call "numbers" at all. Indeed, they are usually presented in the modern form of "axiomatics" that has evolved from the Greek form, the objects and operations being entirely undefined but satisfying certain axioms that embody the associative and distributive laws, for example. Of course they have important interpretations; else they would not be of any use and would not have been introduced. And in these interpretations the objects may be polynomials, functions, matrices, or other mathematical entities that are found to exhibit the properties embodied in the axioms. Such interpretations frequently furnish the motive for setting up the corresponding number systems.

From this it is evident that the modern mathematician has lost the qualms of his forebears regarding the "reality" of a "number" (or other mathematical entity). His criteria for acceptance are of a completely different sort, involving such matters as consistency, utility of the concept, and the like.

---

[20] The parentheses indicate that the summation enclosed by them is to be performed first.

CHAPTER FOUR

The Processes of Evolution

Up to now considerable attention has been paid to historical matters; this was necessary in order to have a basis for analyzing evolutionary forces.[1] Material that a history would certainly be expected to include has been omitted, since all that is needed is sufficient evidence of a historical character to bring out those forces of a cultural nature that influence mathematical evolution. Nevertheless, much detail has been included that was needed for the subsidiary purpose of clarifying the history itself—such as the concept and nature of an infinite decimal. It seems advisable, therefore, before concentrating on the forces involved, to preface the discussion with a brief summary of the pertinent historical facts, omitting most of the details and emphasizing the evolutionary aspects.

In sketching the development of number, and the manner in which geometry affected it, attention was frequently called to the types of stress at work. Without such stresses, there would have been no change in the character of primitive number (not even

[1] Use of the word "force" may not seem appropriate; for example, to call "cultural lag" a "force" appears to strain the current use of the word. Perhaps "stress" would be preferable; it has been used synonymously throughout, it will be noted. Incidentally, Isaac Newton, in his *Principia* (see Cajori, 1934, p. 2, Def. III) defined: "The *vis insita*, or innate force of matter, is a power of resisting, by which every body, as much as in it lies, continues in its present state, whether it be of rest, or of moving uniformly forwards in a straight line."

any primitive number!), nor would mathematics itself have emerged as a special, recognizable element of Western culture. For not only was mathematics dependent on the evolution of the number and geometry concepts, but this evolution was essentially a product of the stresses imposed on it. Ultimately these stresses operated also in the development of mathematics as a whole. In making these stresses, or forces, the main object of study, therefore, the forces that operate to produce mathematics itself are being scrutinized; indeed, the same forces can often be found operating on a much larger scale, namely in the development of all science.

1   THE PRE-GREEK ELEMENTS   So far as what might be called the beginnings of the mainstream of mathematics are concerned, the evidence provided by recent researches of Neugebauer, Thureau-Dangin, Sachs, and others into the Sumerian-Babylonian mathematics has settled many of the questions regarding the origin of numbers and our ways of representing them, and most of the mysticism and mythology connected with them has been cleared away. We refer, for instance, to statements such as the following, which appears in a recently published book for high school teachers (National Council of Teachers of Mathematics 1957, p. 7): "The numbers 1, 2, 3, 4, 5, . . . , are called *natural numbers* because it is generally felt that they have in some philosophical sense a natural existence independent of man. The most complicated of the number systems, by way of contrast, are regarded as *intellectual constructions* of man" (Italics mine: RLW).[2] That is, irrational numbers, complex numbers, and the arbitrary real number are human constructions, while the numbers used in counting are not, but have some kind of special existence of their own. Presumably this feeling is a reflection of the fact that the natural numbers have come down to us, ready-made, from

[2] Compare the nineteenth-century mathematician Kronecker's oft-quoted remark: "The integers were made by God, but everything else is the work of man."

an antiquity most of whose aspects are preserved in folklore rather than in historical documents. However, new historical research enables us to form a much clearer picture of the development of our number system than was formerly available. Moreover, the scientific basis of mathematics has been made apparent, and the reasons for its special place in the theory of knowledge can be more fairly assessed.

In no culture is it known when man first began to count. No doubt he never really ever *began,* in the sense that even if all the relevant historical information were miraculously made available, the development of counting was so gradual as not to allow any determination of a precise time for its beginning (*cf.* Chapter 1, Section 1.1). All cultures apparently had number words, even if only the equivalent of *one, two, many.* But a paucity of number words in a culture did not mean that the bearers of that culture had no way of telling that a collection of ten things has more elements than a collection of nine things, for they could. To do so they may have used devices other than linguistic—such as pebbles, knots in strings, and the like—or by repetition of the elementary number words they made up for deficiency of number words. As pointed out in Chapter 1, there is evidence of the use of mechanical devices, such as tally sticks, even in Paleolithic times (see, e.g., Struik, 1948a, p. 4). As civilization developed, and societies became more complex, necessitating the invention of calendar systems, measuring of areas devoted to agriculture, and the like, the invention of new and more refined means for enumeration, both linguistic and mechanical, became necessary. In particular, this was the case in the old Sumerian culture and in the Babylonian into which it merged. At work throughout all these developments was *cultural stress*—a force that can be considered a kind of "necessity is the mother of invention" cultural factor, roughly analogous to what the biologist calls environmental stress. Here, however, the stress was due more to the cultural than to the physical environment.

To count, number *symbols,* either linguistic or physical, were needed (Chapter 1, Section 1.1); and as the things to be counted became oppressively numerous, and repetition became too burdensome, selection of a base appeared, cipherization accelerated, and a so-called place value system of expressing numbers developed, as it did in the Babylonian culture. With increasing complexity of the social structure, there arose a necessity for introducing operations on numerals, combining them by addition and multiplication for instance; and the invention of the place value system forced the invention of a new symbol, the *zero,* which at first merely indicated the absence of a numeral and not a number in its own right. Whether the zero that we use, which we obtained from India by way of the Arabs, was invented by the Hindus or borrowed by them from the Babylonians, possibly by way of the Greeks, which some maintain, is not too important; for the inevitable ultimate invention of a zero, if not sufficiently proved by consideration of the evolutionary process itself, is confirmed by its appearance in the Mayan culture.

However, even before a symbol for zero appeared in the Babylonian mathematics, calculation with numerals was carried out. Elaborate multiplication tables were constructed. Moreover, it was recognized that division by a number n is equivalent to multiplying by its reciprocal, $1/n;$ extensive tables for reciprocals to be used for division have been discovered. Ingenious methods were apparently worked out for finding reciprocals. It is well to reflect on the great generality and abstraction that had already been achieved by this time. The evolution from the primitive number words that were strictly *adjectival* in nature (*cf.* Chapter 1, Section 1.2e) to the *nouns* one, two, and so on, standing for abstract concepts of number, was long and laborious. Whether the Babylonians went this far, one cannot say, but their manner of using numerals in computation might indicate it, and evidences in their culture of the type of numerology that is associated with mysticism would also indicate it. The common process whereby a symbol originally designed for one purpose is gradually attributed a new significance can be observed in the manner by which the symbol 0 was elevated to be the representative of a *number*. And when, as a numeral, it stood for an abstract number concept, the final step away from the external real world had been taken, and numbers existed as distinct entities. Furthermore, operations with these numerals (addition, multiplication, etc.) became a true science—*a science of numbers*—and just as one calculates and predicts in any of the physical sciences today, so did this ancient Babylonian science of numbers deal with the external world of reality found in the environment of that day.

There were two aspects of this Babylonian science of numbers that were singled out for comment (Chapter 1, Section 3.2), because of their significance for later developments in mathematics. One of these concerned the matter of *theorem* and *proof*. As remarked earlier (Chapter 1, Section 3.2), it is generally considered that the Babylonian mathematics contained nothing that would be called a theorem today; that is, a general statement with logical proof thereof. It was a "do this, do that" kind of affair. Nevertheless, there seems to have been a tacit acceptance of certain unstated rules, as, for example, the processes used for solving equations, which, although only exemplified by instance after instance, would have formed "theorems" if stated in formal fashion. These "mathematicians" clearly knew that certain methods would work, and even though they left no formal statements of them, it is quite conceivable that their mode of communicating them, particularly in teaching, was by *verbal* statement of rules (i.e., "theorems"). And if such was the case, they were close to the notion of a general theorem.

Moreover, for the Babylonians "proof" may have been something to be found in the external world, not in the conceptual world; "proof" may have been purely the demonstration that a theorem—in this case, a method verbally stated—*worked* in example after example, as well as in applications. (Such an interpretation is entirely in accord with what some of the mathematical primitives who manage to enroll in colleges today consider a "proof"!) The view may be taken that the Greek mode of theorem and proof was a natural improvement on that of their Babylonian predecessors, much as our modes are an improvement on those of the Greeks. "Sufficient unto the day is the rigor thereof."

The second noteworthy aspect of the Babylonian science of numbers previously mentioned (Chapter 2, Section 3.1) referred to an operation that seems trivial, but had great significance for the later development of mathematics. This was the fact that one of the applications of the science of numbers was to the measuring of lengths such as the width of a field or the side of a stone block. In this way, the *counting* numbers served also as *measuring* numbers—a natural evolution, since measurement of a length is fundamentally the counting of the number of certain conventional units in that length. Herein lay the germ of what was later to become the real number continuum, and quite possibly one of the reasons why geometry of the Greek type became a part of mathematics. (The need to think visually or in some kind of pattern would likely have forced inclusion of geometry in some form in any case—perhaps in a primitive kind of topology.) Were it not for the fact that measurements became a cultural necessity, thus bringing the science of numbers into identity with geometric forms (perhaps even to the extent of setting up elements of a geometrical algebra; see Neugebauer, 1957, pp. 149-150), geometry might not have become an important part of mathematics as early as it did, especially in the extreme sense in which it dominated Greek mathematics. But, and more important from our point of view, the introduction of the arbitrary real number in the form of "magnitude" was to have a great influence on the way in which the number concept evolved.

The Babylonians devised rules for calculating the areas of rectangles from the lengths of adjacent sides, the circumference and area of a circle from the length of a diameter (with $\pi = 3$), and the volume of a right circular cylinder from the area of the base and its altitude; and they even knew the so-called Pythagorean theorem for suitable integral values of the sides (3, 4, 5; 5, 12, 13; etc.). Although these were just applications of the science of numbers, they formed the material from which *theorems* of geometry are derived. Moreover, they became so well absorbed into the techniques of the professional mathematicians of the day as to orient them toward the status of an integral part of mathematics.

In brief, during the pre-Greek era, *cultural stress* forced the invention of counting processes that, with increasing complexity, necessitated suitable *symbolization*—hence the introduction of numerals. The latter was aided by the fact that the Akkadians took over Sumerian symbols of an ideographic nature, a *diffusion* process, and this in turn led to further *abstraction*. Demands of engineering, architecture, and the like led to application of numbers to geometric measurements and to the gradual assimilation by the number scientist of geometric rules, which were later to become theorems. The number scientist was being subjected to greater and greater stress, from *without* (cultural stress) by the demands of his nonmathematical brethren and from *within* (hereditary stress) by the need for systematizing and simplifying the processes by which he obtained his results of a numerical nature. The symbols themselves began to take on an inherent and mystical meaning, as symbols sometimes do; and number began to emerge as a thing in itself, a concept, divorced from physical reality and applications. The stage was set for the great Greek developments in mathematics.

2   THE GREEK ERA   It has generally been considered that there was a great gap, a kind of "missing link," between the work of the Greeks and their predecessors. There certainly was a gap, but was it of such a nature as to cause the wonderment so often evidenced by historians? It is possible that, when one considers the underlying processes at work, one may arrive at the conclusion that the so-called gap was only the leap usually induced by the introduction of profound revolutionary concepts. The Babylonians had brought mathematics to a stage where two basic concepts of Greek mathematics were ready to be born—the concept of a *theorem* and the concept of *proof*. The ideas of the Babylonians had undoubtedly diffused throughout the eastern Mediterranean area; just how much the Egyptians got from them in this way is problematical, but there is little in Egyptian mathematics, except for other and sometimes more advanced geometric rules, that is not already to be found and superseded in Babylonian mathematics.

Allowing for differences that would accompany adaptation to the ways of a new culture, there is little question that Greek mathematics represented a natural evolution from Babylonian mathematics. A phenomenon that usually occurs when ideas diffuse from one culture to another was here repeated; the "host" culture devised its own modifications of the cultural elements adopted, in order to fit them into its own patterns of thought and behavior. When one contrasts the Greek culture with those of Babylon and Egypt, one is struck by the fundamental difference in their intellectual climates. The Babylonian and Egyptian cultures were, like more primitive cultures, rigidly conformist; independence of thought was not encouraged and would have been frowned upon as a threat to the state. But in the Greek culture, a much freer attitude prevailed; as one author (Barnes, 1965, pp. 122 ff) put it, the birth of a free thought occurred in Greece and what could be called a scientific point of view developed. That this general aspect of the culture had a profound effect on the manner in which Greek mathematics developed (an instance of cultural stress of an environmental character) can hardly be doubted. Nearly all the men of mathematical importance during the first century and a half of Greek mathematics, that is, from about 600 to 450 B.C., were philosophers, as, for instance, Thales, who according to some writers was one of the first to have taken the great leap to theorem and proof, and Pythagoras, whose philosophy and mathematics are so intertwined as to be hardly distinguishable. Although here the lack of historical evidence forces us to conjecture, it is quite possible that it was this wedding of the Greek philosophical bent and the Babylonian science of numbers that caused what is so often termed a gap, but which was actually only a leap to a higher level of abstraction. What more natural, then, that the penchant of the Greek culture for introducing law into the universe should induce a search for the same in mathematics? A discontinuity occurred, to be sure, but it was the kind of discontinuity that might be expected from the standpoint of the evolution of culture. What greater evolutionary gap, indeed, could one hope to find than in the very beginnings of culture itself—the gap between the behavior of man and the behavior of all other forms of life? Man learned to symbolize, and, once possessed of the faculty, he created the gap that we now observe between man's conceptual capacity and the apparent absence thereof in other animals. Similarly, the ability of the Greek philosophers for seeking and finding general laws was just what the mathematics of 600 B.C. needed in order to convert it into a new and more effective instrument, hardly recognizable in the material from which it sprung.[3]

In the opinion of most scholars, mathematics really matured with the Greeks. The concepts of *number* and *form,* which with the Babylonians and Egyptians were so closely tied to external reality, were converted into new concepts on a higher level of abstraction. Although the new concepts were still related to the external world, they achieved a new dignity in that they were presumed to be representative of ideal configurations in an ideally

---

[3] According to A. Szabo, 1960, the Greek deductive method "was built on the Eleatic philosophy," and the first deductive science of the Greeks was the arithmetic. However, he believes that the introduction of axioms and postulates was of geometric origin, induced by the "inconsistencies" inherent in the infinite divisibility of the line segment (Zeno) in contrast to the indivisibility of the unit as they conceived it in their philosophy.

perfect realm of ideas. We have here a curious *duality*. For while geometry, for instance, was considered to give precise descriptions of forms derived from the external real world, it was evident that perfect straight lines, circles, and triangles do not exist in the external world, hence they were presumed to have perfect prototypes in a superhuman ideal world. Thus mathematics, while still *science* in the sense that it furnished a way for dealing with problems of the external world, became something that could be pursued for its own sake, and so was born that dual nature of mathematics that will be discussed further later. In the view of Plato, the scientific aspect of mathematics, that is, its relation to the external world of reality, was of little importance, in comparison with the study of mathematics for its own sake. The history of the period in Greece when discoveries were being made in connection with the solution of basic problems in geometry engenders a feeling that much the same excitement must have prevailed among mathematical researchers of that era as attends the process of mathematical research today, what is today called a real "research atmosphere."

Insofar as *method* is concerned, the greatest concept that evolved in the Greek era was undoubtedly the axiomatic method. The notion of selecting a few concepts as primary, and stating a few axioms or laws concerning them from which to derive all other concepts and properties by means of definition and logical deduction, was to become one of the most important tools for both mathematics and science. True, it was still to undergo further evolution before it could realize its full potentialities, but the big step had been taken. And divorced from its idealistic dual, Greek mathematics constituted a true science, the science of the observable forms of the external world.

It is interesting to observe, too, that before the end of the so-called Greek era, the elements of analytic geometry and the calculus were being worked out. But unfortunately Greek mathematics had already attained its zenith and was on the decline. No doubt the reasons for the decline were in part due to the failure to parallel conceptual achievements with an equally capable advance in symbolism, particularly of an algebraic variety. One can surmise that had the Greeks paralleled their geometric algebra, based on magnitude, with symbolic devices such as were invented by their successors, their mathematics might have found new life. It should not be overlooked that the Greek culture had many numerical aspects, especially in astronomy, where the Babylonian sexagesimal system continued in use. But here the stress that was operating is clear: the sexagesimal system permitted unified expression of both integers and fractions, and so was well adapted to formulating astronomical tables. No such stress, *within* mathematics, was strong enough to force contriving a new symbolism for the solution of problems that Eudoxus and his successors had already handled satisfactorily. It should not be overlooked, too, that after a solution was effected geometrically, there was either cultural lag or cultural resistance, possibly both, involved in the failure of the Greeks to adapt the Babylonian symbolism to their own numerical problems.

But there were undoubtedly other reasons for the decline, *external* to mathematics. Many agree that if the general cultural environment had developed differently, the tradition of Euclid might have evolved in a manner that was not to be exemplified in the evolution of science until nearly sixteen long centuries later. This seems indicated by the work that centered at Alexandria after Euclid. Thus Archimedes, certainly one of the greatest mathematicians, who worked under Euclid's successors at Alexandria, applied his mathematical genius to mechanics, using methods closely allied to those of the modern era, although ultimately using axioms and proceeding by deduction to obtain the necessary theorems; this is both in the Euclidean tradition and prophetic of what was later to be done by Newton. "He invented the whole science of hydrostatics" (Archibald, 1949, p. 23) and worked in astronomy, writing "a book on the construction of a sphere so as to imitate the motion in the heavens of the sun, the moon and the five planets." Archimedes' friend Eratosthenes, "who taught in Alexandria," made a "very accurate determination of the polar circumference of the earth." Indeed, Euclid himself wrote a book on spherical geometry containing propositions designed for use in observational astronomy, as well as a book on optics and one on music. Aristarchus of Samos, who came between Euclid and Archimedes, "was the first to assert that the earth and the other planets (Venus, Mercury, Mars, Jupiter and Saturn) revolved about the sun, thus anticipating Copernicus by seventeen centuries" (Archibald, 1949, p. 22). Despite the extreme geometrical turn that mathematics had taken, the Greeks were seemingly on the path to modern science in both mathematics and its applications.

Even mechanical gadgets had begun to appear, "such as a siphon, a fire engine, a device whereby temple doors are opened by fire on an altar, an altar organ blown by the agency of hand labor or by a windmill" (Archibald, 1949, p. 25); "an automatic machine for sprinkling holy water when a five-drachma coin was inserted" (Kline, 1953, p. 62). According to Kline, steam power was generated and used "to drive automobiles along the city streets in the annual religious parades. When produced by fires maintained under the temple altars, the steam put life into the gods—who raised their hands to bless worshippers; also gods who shed tears and statues that poured out libations";[4] devices certainly calculated to cause grave doubts in the heretically inclined! One can justifiably conclude that it was those cultural stresses, external to mathematics, that came to dominate the course of evolution of the entire Western culture, which were chiefly the cause for the gradual dying out of Hellenic mathematics. And that, as happened later during an era of ingenious mechanical experimentation in France, ideas having great potential "died on the vine" because of

[4] From "Mathematics in Western Culture" by M. Kline, Copyright 1953 by Oxford University Press.

Part of a Greek computer used about 2000 years ago. (Derek J. de Solla Price, from the fragment in the National Archaeological Museum, Athens.)

a lack of demand for them in the cultural environment. To put it another way, science had more than satisfied the demands created by the cultural stresses of the period.

Courant and Robbins state (1941, p. xvi) as their opinion that the ascendency of geometric over quantitative form in what they term "the thicket of pure axiomatic geometry" led to "one of the strange detours of the history of science" and that thereby "a great opportunity was missed." And that "For about two thousand years the weight of Greek geometrical tradition retarded the inevitable evolution of the number concept and of algebraical manipulation, which later formed the basis of modern science." But the hereditary stresses in mathematics that were largely responsible for this "detour" were unfortunately not matched by environmental stresses expressing the need of an analytic geometry and a calculus suitably symbolized. The onus for the two thousand year gap in the development of the number concept and of algebra should not be placed entirely on mathematics, but should be shared by the entire cultural complex of the time. A broad view of Greek mathematics shows that it had an intrinsic vitality and breadth which bore every evidence of proceeding in a direction toward modern mathematics. Not only the course of mathematical evolution, but that of all the intellectual (scientific and humanistic) achievements of the Greeks were to be impeded by the cultural environment and its general decline. Given a different set of environmental circumstances, the algebraic developments, which were later undertaken by the Arabs, might have been created in line with the Greek ways of thought, and modern science might have been developed sixteen centuries ago.

What were the important evolutionary factors in the Greek period? First and most important was the *diffusionary* element, through the agency of which Babylonian-Egyptian mathematics met the Hellenic philosophy and ultimately became an axiomatic and deductive science. *Abstraction* to a higher level, elevating mathematics to the stature of an object of study in its own right, became operative, accompanied by *generalization,* a tool that modern mathematics uses constantly. A new mathematics evolved which, although on the one hand a description of certain structural aspects of the external world, was on the other hand a search for properties of an ideal realm that one might explore with no regard for whether what one found would be exactly duplicated in the external world. In this more abstract atmosphere, internal or *hereditary stresses* played a much greater role than in the Babylonian-Egyptian mathematics. But lack of stresses forcing attention on the role of *symbolism* helped to weaken the advances made, and in the end mathematics and the sciences that it helped to engender succumbed to the stronger external cultural stresses that came to dominate Western culture.

3 THE POST-GREEK AND EUROPEAN DEVELOPMENTS   In comparison with the literary and philosophical work of the Greeks, their mathematics fared relatively well with respect to survival. Although many important treatises were lost to posterity, many of the works of Euclid and other Greek authors were translated by the Arabs and became known to the European cultures. And about the sixteenth century, diffusion of Byzantine Greek manuscripts to the West assisted in the rejuvenation of geometry in Europe. The Arabs also built up an algebraic tradition that was active both in the Eastern Mediterranean region and in the Western Mediterranean area around Spain and Morocco. Mercantile contacts with the Italians led to diffusion of the Arabian algebra, giving rise to an interest in algebra in Italy, especially in the solution of equations—an interest that eventually diffused north and west to find expression in the work of Abel in Norway and Galois in France. The arithmetic of whole numbers and simple fractions was augmented by the admission of other types of numbers through the demands (hereditary stress) created by the solutions of equations and, later, analysis. This is a good example of an important manner in which symbolic processes often affect the evolution of mathematics; to wit, formal manipulations of symbols may force the introduction of new concepts. It will be recalled how operations with a zero symbol, originally invented as only a sign that a digit was absent in the place value expression of numerals—a "symbol for nothing"—ultimately evolved into a symbol for a concept, that is, the *number* zero. Similarly, for a long time the square roots of negative numbers were not admitted to mathematical respectability; the symbol $\sqrt{-1}$ was considered inadmissible as a numeral. Apparently not even admittance to that ideal realm of mathematics which the Greeks had imagined, was to be gained without prior approval by the external world. However, operations with it, guided by the rules of arithmetic, led ultimately to meaningful results. Moreover, a complete basis for algebra and analysis could not be achieved without it. So at long last the symbol was allowed to represent a *number*.

Through the work of Descartes and Fermat there developed, in the seventeenth century, that fusion of algebra and analysis with geometry known today as analytic geometry, whose invention was obviously no historical accident but the result of a long evolution whose beginnings can be discerned in the work of the Greeks. Apollonius, who lived about 200 B.C., wrote "far more than is contained in any of our American textbooks on analytic geometry, so far as conic sections are concerned" (Archibald, 1949, p. 24). Much of the advance that Descartes made, indeed, was owing to the manner in which he arithmetized geometry, introducing also a good algebraic symbolism. Similar observations can be made in regard to the calculus, soon to be established by Newton and Leibniz. That the Greeks had virtually developed certain features of the integral calculus, as well as an approach to the notion of limit in their theory of approximation, is now common knowledge. Moreover, Newton's predecessors had already worked out much of what is taught in the calculus today. In short, the calculus was the result of a long evolution, which required both adequate external stresses and internal developments of a symbolic character before the formal aspects of what we now call the calculus could reach fruition. And it was not left in its final form by Newton and Leibniz; they were the ones who deserve credit for putting the subject into an effective symbolic and algorithmic form, but they left to their successors the polishing work and the furnishing of a proper foundation.[5]

As one reads the history of these developments, one gets the feeling that about this time mathematics was returning to ways not unlike those of the old Babylonian era, albeit of a more sophisti-

[5] The reader cannot be too strongly urged, for a fascinating discussion of the evolution of the calculus, to consult the work of C. Boyer, 1949.

cated character, perhaps. Just as the Babylonians devised rules for computations without justification by what would later be considered sound logical proof, so was much of the seventeenth- and eighteenth-century analysis devised without much more justification than that "it worked." Despite the "strictures of a Bishop Berkeley" and other critics (compare the introductory remarks of Chapter 3), one continued confidently seeking new methods that would give the desired results and did not worry too much about a sound foundation. Actually, there was some concern about lack of rigor—for instance, Newton remarked that one should determine whether infinite series converged—but it was not a time for rigor; it was a time for pioneering in new paths, and boldness was required, not cautious timidity. And one can see the same evolutionary forces at work as in the Babylonian era—both external and internal (hereditary) cultural stresses, with the former dominating, and increasing complexity forcing rapid symbolization without adequate foundation. The diffusionary processes were even more active and subtle, due to better communication and the printing press. And much as the Babylonian era prepared for the great foundations later laid by the Greeks, so seventeenth- and eighteenth-century mathematical analysis was preparing the ground for the foundations of analysis to be laid by such men as Dedekind and Weierstrass in the nineteenth century.

3.1 *Non-Euclidean geometry*   Meanwhile, on the geometric side other developments of great import were taking shape. From the time of Euclid, it was believed that the Greeks had laid down an adequate foundation for mathematics, as embodied in Euclid's *Elements*. We know now that Euclid's axioms were not sufficient—not even sufficient to prove his first theorem. But the feeling of the ancients was quite the reverse, in that they felt he had assumed *too much*. In particular, the so-called parallel axiom was considered unnecessary, and even Euclid is reputed not to have liked its assumption. We have already remarked in the Introduction (Section 3) how for centuries mathematicians tried to prove it as a theorem from the other axioms. Possibly even Omar Khayyam worked on it—there is evidence of this in the work of a fourteenth-century Arab which some feel to be a repetition of Khayyam's work. And we recall that in the first half of the eighteenth century, the Italian Jesuit, Saccheri, worked out a considerable piece of non-Euclidean geometry in the attempt to prove the axiom (which he thought he did). It is not surprising that early in the nineteenth century a breakthrough occurred through the medium of three mathematicians working independently—Gauss, who failed to publish his results, Bolyai, and Lobachevski.

The implications of the discovery that consistent non-Euclidean geometries can be constructed were important for the evolution of human knowledge on all levels, from the most practical to the most abstract (*cf.* Introduction, Section 3). Although it took the mathematical world at least thirty years to come to a realization of the significance of the fact that there was no absolute or necessary character to Euclidean geometry, once the cultural block was overcome, the impact was tremendous (compare Chapter 2, Section 5.1). First, it became evident to all whose minds were not absolutely closed that the Platonic world of ideals in which mathematical truths were supposed to reside must be replaced by a world of mathematical concepts that is man-made and is just as cultural as any of the other systems that man invents in his effort to adapt to and to control his environment. The dual character of mathematics was retained; mathematics was still an instrument for scientific investigation, but on the conceptual side it now achieved a freedom that it did not know before (see Chapter 2, Section 5.2). This was freedom accompanied by a conviction that it was no longer restrained by either an ideal or an external world, but that it could create mathematical concepts without the restrictions that might be imposed by either the world of experience or an ideal world of "truth" to whose nature one was committed to limit his discoveries. This feeling of freedom was not entirely justified, but for the time being it was a grand tonic. (See Chapter 5, Section 1.4.)

Second, a new field was now ready to be opened up and cultivated, namely, the study of axiomatic systems. The traditional point of view was that axioms are self-evident truths. In all the investigation of the axiom of parallels, no one questioned its truth —an interesting type of cultural resistance and probably the principal reason why non-Euclidean geometry was not discovered earlier. Now one had to give up the word "truth" and think of an axiom as simply a basic assumption that one makes and that is descriptive of some model in either the external or the conceptual world. The applicability of the axiomatic method to a wide variety of models, mathematical and nonmathematical, is a consequence of the new conception of the method as it was worked out during the last century and the early part of this century.[6]

3.2 *Introduction of the infinite*   Very likely the new attitude toward mathematics—the feeling that it had cast off its bonds and could now stand on its own feet—was involved in another important development of the latter part of the last century. Today, mathematics is sometimes called the science of the infinite.[7] Until about 1895 such a statement would have had little validity. The Greeks tended to avoid the infinite. Euclid's basic axiom stated that "Every line can be extended"; it did not say that every line was infinitely long. And although it is commonly asserted that he proved the existence of infinitely many prime numbers, what he really stated and proved was that "prime numbers are more than any assigned multitude of prime numbers" (see Chapter 2, Section 3.2); that is, given any finite set of prime numbers, there is a rule

---

[6] The change in point of view was not, to be sure, due exclusively to the non-Euclidean geometries, since it seems to have been part of a more general development. Thus Boole in logic, Peacock in algebra, and Whewell in mechanics were already experimenting with axioms in a manner prophetic of later developments.

[7] H. Weyl; *cf.* Chapter 3, Section 1.1.

by which one can find a prime number not in the set—which is strictly analogous to the axiom about the extendibility of lines. And as was remarked earlier (Chapter 3, Section 1.1), as recently as 1831 the great mathematician Gauss protested vehemently against the use of the infinite in mathematics, stating that "Infinity is merely a way of speaking."

But the cultural stresses exerted by the evolutionary process recognize no such principles; if a concept (no matter how repugnant it is or how much cultural resistance it may encounter) offers a way to the conquering of stubborn problems, it seems that it will ultimately evolve and be accepted. And mathematics had encountered such problems. As all applied mathematicians know, the study of wave motion, in the theory of acoustics, heat, and the like, led to a consideration of series whose terms were expressed trigonometrically, and the study of such series led to questions concerning the foundations of analysis that could be cleared up by a consideration of infinite collections. Eventually this led to the discovery that just as in the case of finite collections two sets may have different numbers of elements, so too in the case of infinite collections two collections may have different "numbers." But for this latter statement to make sense it is, of course, necessary to know what is meant by the "number" of an infinite set. Not only did the German mathematician, Cantor, who studied these problems, give a definition of number for infinite sets—the so-called *transfinite numbers*—but his definition was such that, when applied to finite sets, it yielded the finite natural numbers as special cases. Moreover, he then generalized the arithmetic of the natural numbers to these transfinite numbers. Thus was born the theory of sets. (See Chapter 3.)

Again the mathematical world was faced with the dilemma of whether to admit a new, suspicious character into the realm of orthodox mathematics. Evidently the newfound freedom engendered by the non-Euclidean geometries was not, when one came right down to it, to be considered as a license to introduce concepts too far divorced from what one called reality. After all, one could easily make out a case for the applicability of non-Euclidean geometry to the physical world. But where in the external world was there an infinite set? It is not surprising that Cantor's most fundamental and history-making paper was at first denied publication by the journal to which he submitted it. And although his papers on the infinite were finally published, the reception accorded them was by no means cordial.

Nevertheless, the new theory opened up new vistas for investigation, in addition to offering a way of solving troublesome fundamental problems, and the resulting hereditary stress proved too strong for the opposing cultural resistance. By 1900 the theory had essentially achieved respectability in mathematical circles, although it still was not universally accepted (and in the meantime its sensitive creator, Georg Cantor, had been driven to a mental institution).

4 THE FORCES OF MATHEMATICAL EVOLUTION The principal forces discernible in the development of mathematics are listed below. Most of them have already been observed in the previous discussion, but not all, since some have become noticeably effective only in modern times:

1. Environmental stress
    (*a*) Physical
    (*b*) Cultural
2. Hereditary stress
3. Symbolization
4. Diffusion
5. Abstraction
6. Generalization
7. Consolidation
8. Diversification
9. Cultural lag
10. Cultural resistance
11. Selection

4.1 *Commentary and definitions* Much as in the case

of the development of the individual man, where two main influences are recognized—the *environmental* and the *hereditary*—the flow of mathematical evolution is influenced by external and internal stresses. Borrowing from this analogy, the same terms, *environmental* and *hereditary,* have been used to differentiate between these types of stress. However, the mistake of thinking that thereby any neat separation has been effected must be avoided, for in the evolution of a given mathematical concept both kinds of factors will usually be found at work and, as in the case of natural evolution, are difficult to disentangle. For purposes of analysis, nonetheless, it is convenient to make the separation, always with the realization that it may be impossible in any given situation to effect it in actuality.

The environmental stress is separable into physical and cultural components. However, the physical component has been of importance chiefly in the inception of counting—the "one-two" stage (see Chapter 1, Section 1.1)—while in the extension to a true counting process (and most later developments) the cultural component was dominant. One should not be misled by the fact that physics and mechanics were a major factor in the development of mathematics. Physics, like mathematics, is a cultural phenomenon—part of the cultural environment that man has built. And while the physical component of environmental stress may continue to be one of the most important influences on the evolution of *physics,* this is not the case with mathematics.

In the Hellenic culture the influence of hereditary stress becomes quite noticeable. It was evidenced most strongly in the "crisis" due to the discovery of incommensurability and the space-time paradoxes of Zeno. These were no doubt a major factor in the introduction of both the axiomatic method and the development of geometry as a tool for investigating number theory, as well as for the study of spatial forms such as the triangle and circle. Not to be overlooked also is the cultural stress exerted by the general Greek philosophical outlook, with its attendant desire for knowledge of the fundamental structure of the universe; this, too, played a role in the development of Greek geometry as it is presented in the *Elements* of Euclid.

Symbolization was basic in the development of counting and ultimately led to special types of symbols for mathematics of an ideographic nature. So long as one was tied to the natural language, the language of common discourse, or even special *words* created expressly for mathematical purposes, mathematical advance was hampered. (Compare the remarks on cipherization in Chapter 1, Section 2.2.) The tremendous spurt in mathematical analysis during the seventeenth century was not, it seems fair to say, merely a concomitant of the general cultural advances being made in Europe at that time, unless, indeed, one manages to tie these advances in with the great symbolic achievements of Vieta, Descartes, Leibniz, and others of the period. For on analyzing the mathematical progress of that time, one is struck by how much it actually consisted of the invention of a new and powerful symbolic apparatus. Although in the case of the calculus, for example, beginnings are discernible in the work of Archimedes and other Greek mathematicians and by the time of Newton and Leibniz an extensive theory of integration and differentiation had already been worked out, it was Newton and Leibniz who created for it "a general symbolic systematic method of analytic operations, to be performed by strictly formal rules, independent of the geometric meaning . . . it was just this Calculus which was established by Newton and Leibniz, independent of each other and using different types of symbolism" (Rosenthal, 1951).

The function of a special symbolism in mathematics can be compared to the function served by habit in our daily activities. We need not think through the processes of tying our shoes, for instance; we have developed a habit that does it for us. Similarly, solving a quadratic equation takes no thought once the formula embodying the solution has been memorized; we have developed a symbolic "habit" to do it for us. It is unfortunate that so many mathematics students develop only such symbolic habits and know so little of their background. Indeed, many mathematics teachers have had the experience of receiving bitter complaints when their teaching goes beyond the instilling of symbolic habits (*cf.* Introduction, Section 2). Moreover, as remarked earlier, once a suitable symbolism is set up, it often creates of itself an internal stress of hereditary character,[8] such as the suggestion of a generalized form, or the need of an extended theory for the purposes of interpretation (e.g., the stress exerted by higher-degree equations toward extension of the notion of number to include $\sqrt{-1}$ as a number).

The initial impetus for the great advance made by the Greek culture in mathematics was given by *diffusion,* whereby Babylonian and Egyptian mathematics met Greek philosophy and produced a fusion that was a new and entirely different kind of mathematics.[9] During the preceding period, when geometry existed only in rules for calculating areas and volumes, the same type of cultural stress that produced the science of numbers also produced the rules of mensuration. In addition, hereditary stress led the practitioners of the science to make up problems of a geometric nature for the sole purpose of exercising their arithmetic powers. But besides these types of stress, very little evidence is found of other evolutionary factors. And without new cultural contacts, arithmetic and geometry of the Babylonian-Egyptian type could well have remained in the virtually static state in which Chinese mathematics found itself. Diffusion from Babylon and Egypt to Greece provided new incentive for the Greek philosophy. Even those histories that accept the folklore regarding Thales and Pythagoras have these

[8] This was undoubtedly what Hertz felt; *cf.* his statement quoted in Section 3 of the Introduction.

[9] During the years 1930-1945, roughly, diffusion from the mathematical centers of Germany and Poland led to the creation of much new mathematics. When Hitler expelled Jewish mathematicians, and incidentally influenced many of the best so-called "Aryan" mathematicians to leave with them, he may have done irreparable damage; but there was a compensating factor in that so many European mathematicians, Jewish and non-Jewish, were led to mix intellectually with their colleagues in this country.

gentlemen engaging in widespread travels through the Near East and gleaning information of a geometric nature as a basis for their reflections. According to S. Gandz (1948, p. 13), "It is quite safe to say that the propositions of Euclid's *Elements* II 1-10 are expounding theorems of Babylonian algebra in geometric form."

Comparison with the number science of their predecessors reveals also that the Greeks introduced new components into the process of mathematical evolution, especially those of *abstraction* and *generalization*. To be sure, this is a matter of degree—these did not *start* with the Greeks, any more than an individual person started the counting process. The development of the primitive concepts of number and length (as well as of standards of measurement) in the Babylonian and Egyptian cultures had already stimulated some abstraction and generalization of an elementary kind. And, indeed, the same can be stated of the symbolization that underlay these achievements. But it is essentially with the Greek development of mathematics that one finds the peculiarly mathematical form of abstraction and generalization that is familiar to every modern mathematician. The pre-Hellenic types of abstraction and generalization were somewhat analogous to those employed by the modern engineer when he sets up suitable mathematical models to fit a given "real life" situation. The latter might be called a "first-order" type of abstraction and generalization, whereas the Greeks introduced a "second-order" type, in that it built upon the "first-order" elements already existent in number science and rules of mensuration.

It seems unnecessary to comment further on the processes of abstraction and genralization. From his contacts with physics and other fields of science, including the social sciences, the mathematician abstracts mathematical models. The scientific character of mathematics is chiefly due to this. And he abstracts constantly (a "second-order" type of abstraction) under the influence of hereditary stresses—as, for instance, when he uses the axiomatic method to study the intrinsic properties of a concept that he has discovered to be present in a number of different mathematical theories. The stress in such a case is somewhat of an "economic" nature, in that one must economize on his time by working out the properties of a concept once for all, and not over and over again in each case where it makes its appearance in a special guise.

The term *consolidation* is used to indicate the process by which diverse and scattered mathematical systems are brought together and encompassed in one system. In some cases it may consist merely of a fusion of two systems, combining the advantages of both. For example, the use of the Ionian ciphers along with the Babylonian sexagesimal place value system by Ptolemy and other astronomers was a type of consolidation (persisting, as has already been remarked, to our own day—only with our own numerals used instead of the Ionian). The union of algebra and geometry, to form analytic geometry, is another example. Consolidation is usually achieved by the operation of other evolutionary forces, particularly hereditary stress, abstraction, and generalization. It is possible, for instance, that consolidation occurred in some of the primitive development of number adjectives. In cultures that employed varying types of numerals for different categories of objects, the ultimate transition to a single form of numeral for all categories may have been a case of consolidation involving cultural stress or an elementary form of abstraction.

Consolidation has been of more importance during the modern era, however. As mathematics grows, there is not only more opportunity for consolidation, but hereditary stress frequently forces it to occur. Mathematics has long since grown beyond the comprehension of any one individual; and if it becomes evident that theories that have developed in apparent independence of one another do, however, exhibit analogous properties then abstraction and generalization will usually produce a system in which these properties become special cases. It was in this way, for instance, that group theory was born. Today all mathematicians are familiar with group theory and immediately recognize group-theoretical elements in their work, and are thus able to appeal to the well-known (and already worked-out) properties of groups embodied in the theorems of group theory. Until the consolidation of these group-theoretic features of various theories was accomplished, however, the elements of the theory were scattered throughout algbra, geometry, and analysis in different guises.

This brings out again the difficulty in making a neat separation between evolutionary forces; usually one is accompanied or preceded by others. Consolidation is today almost universally caused by hereditary stress and accompanied or accomplished by both generalization and abstraction. The late E. H. Moore is reported by Bell (1945, p. 539) to have stated in 1906: "We lay down a fundamental principle of generalization by abstraction: *The existence of analogies between central features of various theories implies the existence of a general theory which underlies the particular theories and unifies them with respect to those central features. . . .*"[10] Clearly this was a recognition of the consolidation process in its modern form; it might be added to Moore's observation, however, that the "existence" he mentions is only potential until it is actualized under hereditary stress. Instances such as Moore had in mind may be found frequently in the evolution of modern algebra, where recognition of similar patterns in different systems has led to consolidation. Thus all common number systems, including both that of ordinary arithmetic (of integers) as well as the real number system, have aspects which are special cases of a general type of number system now known in Algebra as a *ring*. And in the modern field of mathematics known by the name of topology, the growing abundance of various types of spaces led to the notion of "topological space," a framework within which different kinds of spaces could be identified by addition of appropriate axioms.

Sometimes what appears to be consolidation is actually only

---

[10] From "The Development of Mathematics," by E. T. Bell. Copyright 1945 by E. T. Bell. Used by permission of McGraw-Hill Book Company.

generalization. If, for example, a set of axioms is weakened by deletion of one or more of them, the resulting set may turn out to embody a significant theory common to several theories. But the *process* whereby it was produced was a rather trivial kind of generalization, and not the result of consolidating elements that have been detected as being common to the several theories. The *result* may be the same, but not the processes by which it was attained.

Diversification occurs when, starting from different aspects of a mathematical system, new systems are created that generalize or extend those aspects. Like consolidation, it is of more importance in modern times than it was earlier. However, by taking some liberty with historical details (which do not uniformly fit into the patterns), the proliferation of number systems, geometries, and so on, now existent, has been due in considerable part to diversification. The prime aspect—the *raison d'être*—of the natural numbers was their role in the counting process. Eventually they acquired both measuring and ordering aspects. The operations of addition and multiplication were introduced, because of the demands made by cultural stress, forming an additional aspect. Eventually the operational and measuring aspects led to fractions and, ultimately, to the real numbers; the primitive counting aspect was extended to the transfinite cardinal numbers and the ordering aspect to the transfinite ordinals. The professional mathematician will have no trouble eliciting manifold examples of diversification. Here again abstraction and generalization play a major role, although hereditary stress is usually instrumental in initiating the diversification.

Cultural lag and *cultural resistance,* as general cultural forces, were discussed in Section 2 of Preliminary Notions. Under the former may be placed the part that "tradition" plays, preventing the adoption of a plainly more efficient tool or concept; the metric system of measurement in the United States was cited as a good example in the nonmathematical area. It may be seen at work in both mathematics and mathematical education. No doubt, a kind of inertia is frequently its basis, rather than tradition. Failure of improvements in the symbolization of numbers to diffuse from one culture to another was undoubtedly due, in large part, to cultural lag. Under cultural resistance may be gathered those less passive forces that resist change; it may take the form of nationalism (as in the well-known adherence to the Newtonian theory of "fluxions" in England, in opposition to the Continental form of the calculus), "cliquism" (mathematicians, too, are "human" and will sometimes adhere to terminological or conceptual devices because they may dislike the source of obviously superior counterparts), and the like.

With the passing of time one frequently observes that various symbolic devices (or merely special symbols) emerge for the expression or handling of a concept, with the eventual survival of only one. This is an elementary example of the way in which *selection* works in mathematical evolution. The same sort of thing occurs when various alternative concepts evolve, all directed toward the same mathematical objective, but with the eventual survival of one, although sometimes, when none is plainly superior in all characteristics to the others, several may survive.

Not always, however, is survival a case of superior efficiency. It may be due to such a trivial cause as the dominance of a particular coterie of mathematicians within a culture. This applies to mathematical theory itself as well as to such relatively minor matters as the acceptance or rejection of a particular symbol. This is not to be interpreted as implying overt rejection by a dominant group, however. The term "selection" is perhaps not too fortunate in this respect, since the process is not overt but gradual. A number of years ago a special symposium, attended by an international group of mathematicians, was held for the purpose of selecting— in overt fashion—a standard terminology for a mathematical field in which the terminology then current was quite chaotic and confused. Although many benefits ensued from the affair—as is usually the case when opportunity is afforded for intimate exchange of views—it was a wretched failure so far as accomplishing its original purpose is concerned. The terminology in the field ultimately became standardized, but through the processes of "natural" selection—new workers coming into the field gradually settled on a term here and there, and, except for a few diehards, the problem of selection was finally solved. This is, incidentally, not an uncommon feature in rapidly growing areas of mathematics; various individuals invent terminology whose survival is greatly uncertain until the decision is made by gradual selection.

There are also classic cases, just as in other fields of science, of work that has failed of recognition because the originators were not sufficiently in contact with important mathematical centers. As an organism that forms part of a general cultural continuum, mathematics inevitably grows in the directions taken by those portions of the mathematical community that maintain the strongest ties with the culture; and these are generally to be found in the so-called "important" mathematical centers. In the extreme case of an isolated worker who fails to keep in touch with these centers, perhaps ignoring or otherwise unaware of their publications, there develop what might be called cultural singularities—creations that seem "out of step" with the times. The survival value of these creations may be small; or they may be of great significance and ultimately termed "before their times." In the latter case, if the direction of mathematical research is such as to lead to similar ideas, they will be either "rediscovered" or unearthed by someone who recognizes their significance. (The cases of Grassmann and Gibbs, or, in botany, Mendel, furnish good examples.)

It would be interesting to make a thorough study of the operation of selection factors in the evolution of mathematics; so far as we know, this has not been done. In this connection, attention might be paid to evidences of *cyclic phenomena* and *mathematical fashions*.

A very important type of selection in mathematics, however, is that which controls the direction taken by research, that is, which directs the creation of new mathematics, and particularly the selection of problems that the mathematical fraternity considers "important." This type of selection is unquestionably dominated by all the other forces of evolution, particularly hereditary stress, and is unquestionably deserving of a special study. In time of national emergency, for example, not only may research be redirected toward problems that have been "unpopular" or otherwise neglected, but entirely new branches of mathematics may be created. (For a discussion of selection on the part of the *individual,* the reader is referred to Hadamard, 1949, Chapter 9.)

4.2 *The individual level*   Since the concern here has been only with the evolution of mathematics as a *cultural* organism, those forces that operate on the individual or psychological level have not been discussed. Much has been written concerning "serendipity" in the experimental sciences, that is, fortuitous discovery or invention made during the course of a study or search for something unrelated. The discovery of penicillin is usually cited as the classical example. A reading of the history of modern mathematics reveals analogous cases in mathematics. But serendipidity is not included here as an evolutionary force, since its influence is of much the same accidental character as is the personality of the individual mathematician. Undeniably such factors do influence the course of mathematical development; mathematics grows only through the endeavors of individual mathematicians, even though they are guided and limited by cultural forces. But such matters belong more to the psychological aspect of the development of mathematics. It might be profitable, of course, to make inquiry into the interaction of such psychological factors with the evolutionary forces that have been listed here on the cultural level. Indeed, an inquiry of this kind could conceivably be as fruitful as the study of gene mutation relative to biological evolution.

It might also be asked why mysticism has not been included in the above list of evolutionary forces, since so much prominence was accorded it in the evolution of the concept of number. It seems preferable, however, to consider mysticism as only a special form of cultural stress that played a role in the development of the number concept, rather than as a general evolutionary force; it is not, like the other forces, conspicuously operative in modern mathematics. On the other hand, its influence on the early evolution of number was sufficiently dominating to merit its being considered as forming a "stage" in that evolution. It should not be confused with idealism of the Platonic variety; many modern mathematicians who can hardly be called mystical adhere to an idealistic philosophy of mathematics.

5 STAGES IN THE EVOLUTION OF NUMBER   To avoid the mathematical technicalities that a complete study, replete with cases, would involve, our discussion of evolutionary forces has been limited chiefly to the evolution of number and elementary geometry. These forces have been influential, however, throughout the broad range of mathematical development within Western culture. It must be emphasized that although our approach has proceeded from a historical point of view, the list of forces is not to be considered as constituting a historical sequence. All of the forces listed are usually operative simultaneously—they are, indeed, presently operative.

By way of contrast, a list is given below of the historical sequence of *stages* through which the number concept passed on its way to its modern status:

Stages in the evolution of number

One-two differentiation
One-two-many
Comparison of sets of objects (one-to-one correspondence)
Tallying
Number words
Ideographs
Mysticism
Numeral systems
Operations with numerals
Idealism
New number types (complex, real, transfinite, etc.)
Logical definition and analysis

The last stage—logical definition and analysis of the number concept—has not been discussed, since it requires knowledge of too many technicalities. Suffice it to comment that, chiefly due to hereditary stress, the task of affording an acceptable definition of number, exhibiting all the traits that the intuitive concept would lead one to expect of such a definition, has been taken up by the modern school of logic and foundations of mathematics. A number of such definitions have been formulated, generally on some kind of axiomatic basis for the theory of sets. Mathematicians generally, however, get along quite well with the intuitive conception of number, since it seems adequate for the purposes of the "working mathematician."

CHAPTER FIVE

Evolutionary Aspects of Modern Mathematics

1 MATHEMATICS AND ITS RELATION TO THE OTHER SCIENCES
It is a seemingly paradoxical feature of the development of science that the further its concepts recede from external reality, the more successful do they become in the control of man's environment. Consider physics, for example. Its concepts have become so abstract that years of training are necessary for an appreciation of them; and when one eventually comes to feel that he has an understanding of them, he may have to adopt toward them an attitude very different from that with which he regards the materially perceptive objects of his environment. But modern physics "works"; no matter how abstract and "unreal" its concepts have become, they have enabled us to attain the threshold of what promises to be a new revolution—that of the atomic age. It is difficult, if not impossible, for the participant in great social changes to realize their significance. But one need not have much imagination to sense that, barring catastrophic wars, man has at his fingertips new sources of energy that will change his entire ways of life.

1.1 *Relation to physics*   The relationship of mathematics to physics has been, and continues to be, very intimate. Physics has been one of the most important sources of cultural stress on mathematics, particularly during the last few centuries. However, this has not been a one-way relationship; although many features of classical mathematics are derived from consideration of physical theories, the reverse also holds. One of the most interesting phenomena observable in the relationship between the two fields has been the manner in which mathematical theory has at times advanced far beyond the needs of physics, in directions of no interest to the physicist, only to have the latter ultimately find in the newly created theories of mathematics precisely the tools he needed to revise or extend his own conceptual framework. Mathematical concepts, created originally by abstraction from natural or cultural phenomena, subsequently develop under the influence of evolutionary forces within mathematics until they evolve into forms that suggest new patterns into which to fit natural and cultural phenomena or that provide tools for the study thereof. It is inherent in the nature of evolutionary forces that they tend to induce, ultimately, conceptual structures of cultural significance. It would be interesting to study in detail the evolution of one of the cases in which a modern mathematical theory, whose immediate origin was far removed from external physical reality, found application in a physical theory. This would involve tracing backward, in either or both mathematics and physics, the threads connecting the two theories to their origins in order to see if they have common meeting points. One might expect that such would be the case, especially since the analogous phenomenon can be found *within* mathematics.

Within mathematics, diversification occurs under the influence of abstraction, generalization, and hereditary stress. New types of number systems, new types of geometry, and new algebraic theories evolve, seemingly unrelated to one another. Ultimately, various kinds of consolidation occur. These may take the form of a fusion of an algebraic system and a geometric system as in the classical case of analytic geometry. In modern times consolidation between algebra and a new form of geometry, *topology,* occurred. Not only did this have a profound influence on the future evolution of topology, but the new system, *algebraic topology,* suggested new algebraic concepts that had a strong influence on the development of modern algebra. This is not just a modern phenomenon. When the inadequacy of the earlier forms of Greek geometry was revealed by the discovery that not every two line segments had lengths whose ratio could be expressed in integers (incommensurability), the theory of Eudoxus solved the difficulty by a fusion between number and geometry (*cf.* Chapter 2). In each of these cases, concepts from one area of mathematics were found applicable to another, and a form of consolidation occurred. Similarly, application of a mathematical theory to aid in producing a new physical theory is also a form of consolidation.

Although physics and mathematics form separate disciplines in the universities, the separation is not any more clear-cut than that between certain fields of mathematics. Some thirty years ago, a story appeared in a popular magazine (which one, I cannot recall) in which the "hero" was head of the department of topology in a fictional university. The idea of there being, analogous to departments of physics and mathematics—the form in which the separation of physics and mathematics is ordinarily effected—a separation of mathematics so fine as to result in a Department of Topology was quite amusing to the mathematicians who chanced to read the story. But it turns out that the idea was not so farfetched after all. In some institutions, because of the growth of mathematics and the increasing student enrollment, statistics, actuarial science, and logic have split off from the parent department of mathematics to form departments of their own. Splits between "pure" and "applied" mathematics occurred in some universities at least as far back as the year 1920. On the other hand, collaboration between physicists and mathematicians, ignoring departmental lines, has not been uncommon. For example, graph theory, a branch of topology that had its origin in studies of physical problems during the 1800's and had become rather quiescent during the present century after a period of rapid growth, recently experienced a rejuvenation resulting in newfound applications in both the physical and the social sciences.

1.2 *Tendencies toward greater abstraction in science*

These matters are part of a tendency that is best considered from a broader viewpoint. During the past century, all science has become more abstract. Physics, today, is quite as abstract as mathematics was during the nineteenth century, and theoretical physics is as abstract as modern mathematics. The *orientation* of physics is presumably toward "physical reality"—that is, toward an explanation of physical phenomena. But this does not gainsay the fact that much theoretical physics is conceptually as far removed from "reality" as the most abstract mathematics. Likewise, comparison of the current state of chemical and biological sciences with that of a century ago reveals a similar tendency toward abstraction, especially in their most fundamental aspects. The social sciences, comparative newcomers on the scientific scene, are still in the early stages of their evolution; the data-gathering stage is already giving way to the stage of setting up general theories, and new mathematical tools (statistics, graph theory, linear algebra, topology) are being adapted to their needs. As an evolutionary force, abstraction is clearly not peculiar to methematics.

Moreover, both cultural lag and cultural resistance appear to be even more influential in the evolution of the social sciences than in mathematics. It may turn out to have been one of the greatest misfortunes for the human race that man has been so reluctant to study his own behavior. The types of roadblock thrown in the way of the physical scientist by his cultural environment, during the early development of modern astronomy and physics, have been set up and never completely removed for the social scientist. The modern analogues of the primitive dogmas "explaining" man's origin and behavior are in some respects quite as antagonistic to theoretical advances in the social sciences as were their prototypes in medieval days toward physical theories. And, regrettably, sometimes scientists, unfamiliar with the history of science, assume a scornful and unsympathetic attitude toward the struggles for recognition of the social scientists. This is both unfortunate and unjustified. The same scientist who (viewing the manner in which the achievements of physical science have been utilized to threaten mankind's extinction) bewails the fact that "we" know so little about cultural evolution that we seem utterly unable to prevent the catastrophe, nevertheless looks with contempt upon the only part of science that has tried to do something about our plight. And it may turn out that the effort was too little and too late. However, assuming that the present cultural situation continues to exist, it appears inevitable that the social sciences will achieve a theoretical status that will "explain" man's behavior, individually and collectively, in as efficient a manner as that in which the physical sciences now explain "physical reality."

1.3 *Relation to other sciences in general*   As an evolutionary force, abstraction is just as natural, and fundamental, as any of the other forces listed in Chapter 4. And as the oldest member of the scientific family, mathematics has been longest subjected to its influences. It is hardly to be wondered at, then, that mathematics, the oldest branch of science, should have achieved such a state of abstraction. But is modern mathematics "taking a wrong turning" as some feel Greek mathematics did, and should more notice be taken of the physical environment that formed the initial stress that started mathematics and gave it nurture? To do so would be to be blind to the fact that the major environmental stress on mathematics has, for centuries now, been cultural, not physical—specifically, the needs of sister sciences. And the latter themselves have long since left the practical arts stage and are themselves already abstractions. The high degree of abstraction attained by modern mathematics can hardly be termed a wrong turning; it is the natural product of evolution, as the broad picture of scientific tendencies reveals. It can be expected, however, that mathematics will be ever alert to the needs of other parts of science. Even diversification feeds not alone on the potential of the individual branch of science, but on the neighboring branches.

The admonition to maintain contact with "applications" is actually a part of the larger picture in which the workers in a given branch of mathematics should be continually aware of the possibility for sharing ideas, and deriving stimulus from, other branches of science, *be they within mathematics or without.* Such forces as cultural stress, consolidation, and diffusion of concepts can occur relative to any two or more branches of science, be they within the same general "discipline" or not. For nonscientists (e.g., a governmental agency having funds to dispense for research) to ask that a scientist confine himself to what seems "practical" is to be oblivious of the fact that some of the most abstract creations turn out, ultimately, to have what the average person thinks of as "applications." The cases of Faraday and his researches in electricity and magnetism (making possible the electric motor) and of Clerk Maxwell and his equations (revealing the existence of radio waves) are classical instances. These are matched by the history of mathematical logic—the utmost in abstraction, one might say—and its ultimate importance in the computing industry (von Neumann, creator of ENIAC, was originally a worker in the foundations of mathematics and well acquainted with mathematical logic).[1] However, such "applications" are incidental and only part of the broad scene in which mathematical and, more generally, scientific abstractions diffuse from one part of science to another.

1.4 *Specialization*   All the evolutionary forces listed in Chapter 4 are constantly influencing the current trends in mathematics (with the environmental stress being, as stated above, of a cultural character). Looked at from a broad viewpoint, modern mathematics exhibits a kind of continuous creation process with new branches, such as computer and automata theory, appearing;

---

[1] Like the cases cited in Section 3 of the Introduction, mathematical logic evolved from hereditary stresses within mathematics, not with an eye on "applications."

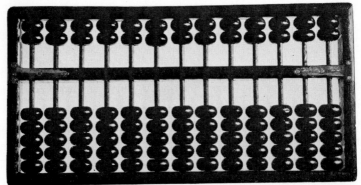

(Top) An Abacus; an ancient form of computing machine which is still in use today. (Fundamental Photos.)

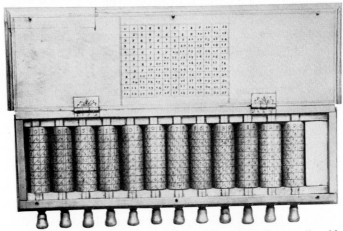

(Middle) "Napier's bones," a device for multiplying attributable to John Napier and used widely in Europe in the 17th century. (The Bettman Archive.)

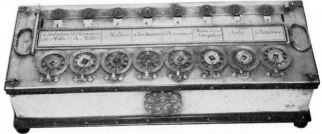

(Bottom) Pascal's adding machine, invented by Blaise Pascal in the 17th century. (Courtesy of I. B. M.)

188

older but by no means ancient fields, such as mathematical logic and topology, already in mature stages; still older fields, such as analysis, being continually invigorated with the aid of such new fields as topology and set theory; and, finally, really ancient fields, such as number theory and classical geometries, hoary with age but by no means dead. It should be evident, from this picture, how impossible it is for any mathematician today to familiarize himself with the entire body of mathematics; life is simply too short. And as a result of the great diversification, it has been necessary to resort to specialization. Both the graduate curricula and the requirements for the Ph.D. degree in mathematics are continually revised so as to provide for as broad a training as possible, while permitting the specialization necessary to provide a basis for the research requirements of the degree. Many decry specialization, but we are faced with an unavoidable fact: when a field becomes very large and varied, limited human resources compel "retrenchment," and specialization is the only way in which progress can be made. This is just one of "the facts of life." It is the mathematical analogue of the extreme specialization in occupations to be observed in the more advanced modern-day cultures. And just as a specialist in any occupation must to some extent keep abreast of the developments in his culture that affect him—political, financial, mechanical, and the like—so must the specialist in a branch of mathematics spend some time familiarizing himself with developments in other branches (as well as his own). In this way, the forces of evolution are brought into play. Specialization is a natural compromise between the great diversification of modern mathematics and the limited capabilities of the human mind. And it has become an increasingly important factor in hereditary stress; it promotes consolidation, for instance, as a means that may enable the individual to command a wider range of conceptual materials.

1.5 *Pure and applied mathematics*   Another effect of the great diversification and the resultant specialization, not only in mathematics but across the entire broad spectrum of science, has been the creation of a speciality called applied mathematics. (See Section 3 of the Introduction.) There has been much misunderstanding and confusion as to what the term means, because of the absence of any generally accepted definition.

In this connection, it may be useful to consider the following schematic diagram:

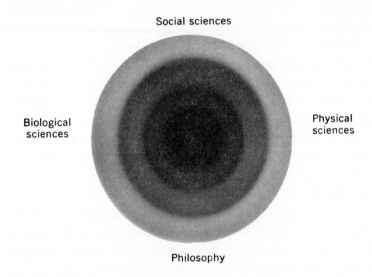

The shaded portion represents mathematics, the heavily shaded part denoting the branches that are the core, so to speak, of modern mathematics. The core may be considered the heart of mathematics—the "purest"—in which the development of the subject per se is chiefly centered. The more lightly shaded portions represent areas of mathematics that have more or less contact with the other sciences, that nearest the core having the least contact and the least-shaded portion having the greatest contact. The unshaded exterior represents, as indicated, the physical, biological, and social sciences, as well as philosophy (which has been strongly influenced by mathematical ideas). The reason for not making any abrupt contrast between the shaded and unshaded areas is to emphasize that in practice there is no abrupt dividing line. A physicist may find himself doing "pure" mathematics, and a mathematician may at times be doing physics. Even though a university may have departments labeled "Pure Mathematics" and "Applied Mathematics" (as some universities do), it does not follow that the backgrounds, especially the training in mathematics, of the members of the two departments can be sharply differentiated. The philosophy behind the departmental separation is chiefly that a member of the "applied" department may be expected to emphasize the study of mathematical concepts that are directly pertinent to the other sciences. In addition to a background in the hard core of mathematics, he must have some familiarity with other sciences, especially with their problems and methods, and the ability to acquaint himself with unfamiliar situations. He may be as well qualified to work in pure mathematics as any of his colleagues in the "pure" department—this frequently is the case —but his interests are different.

On the other hand, the "pure" mathematician—one who works in the heavily shaded area—may, although he has no interest in other sciences, unwittingly create new concepts that the applied mathematician recognizes as useful for the other sciences. It is not uncommon, either, to find the "applied" mathematician developing concepts that his general interests suggest and that turn out to form an extension of the core of mathematics.

The evolutionary forces at work here should by now be fairly evident. Cultural stress exerted by different specialities on one another is foremost in evidence; but the process also involves diffusion from one specialty to another, consolidation followed by further diversification (and conversely), abstraction, and generalization—it is little wonder that labels such as "pure" and "applied" are frequently difficult to apply to the individual mathematician or piece of mathematics. The pure mathematician, the one who, as we say, does "mathematics for its own sake," having no regard whatsoever as to whether what he creates will ever have any applications in what most people call the "real world," is continually being surprised at finding his concepts put to work in this so-called real world in a way that he, the pure mathematician, never dreamed. To put it another way, it appears that no matter how abstract and seemingly removed from physicial reality mathematics may become, it *works*—it can be applied either directly or indirectly to "real" situations—as witness radio, air travel, and the like, none of which would have been possible without mathematics. But this sort of phenomenon only bears witness to the cultural and scientific nature of mathematics. After all, a major function of culture, especially of the scientific component of it, is to adapt to and control man's environment. And although the dual nature of mathematics may seem to split it into a part that can be "applied," as we say, and a part that is just something for professional mathematicians to "play with," there is actually no clear separation. Both aspects of mathematics serve a scientific function, and if the so-called "pure" part that ordinarily functions in the *conceptual* part of the world of reality often becomes instrumental in dealing with the *physical* environment, there should be no cause for wonder. Both it and the conceptual world in which it ordinarily abides trace their ancestry to a world in which abstraction was only beginning to yield, under the play of cultural stress, the concepts that became mathematics.

Nevertheless, the impression persists that while "applied" mathematics serves a useful function for the urgent affairs of life, "pure" mathematics is an "ivory tower" endeavor, having only an aesthetic function. There is little question that "pure" or any other kind of mathematics does give to its devotees an aesthetic satisfaction; as a matter of fact, this may be and probably is the sole reason why most of them pursue it. But it does not follow that this is the only function that it serves, and from the viewpoint of the cultural nature of mathematics it does have another function, namely a *scientific* function.

An opinion expressed by d'Abro (1951, Vol. I, pp. 119-120), who wrote a popular two-volume work entitled *The Rise of the New Physics* is of interest in this connection: "The accepted distinction between pure and applied mathematics is far from satisfying. In the first place, it permits no permanent classification. As an example, let us consider the doctrine of classical mechanics. When the basic postulates of classical mechanics were established by Galileo and Newton, they were thought to express physical characteristics of the world. Classical mechanics was thus regarded as a branch of applied mathematics. But today, as a result of the theory of relativity, we know that the classical postulates do not correspond to physical reality. Strictly speaking, we should therefore reverse our former stand and view classical mechanics as an abstract doctrine pertaining to pure mathematics."[2]

In brief, what is considered "applied" mathematics today may, by a curious reversal of the usual process, become "pure" mathematics tomorrow. And, at any given moment of time, there is no clear distinction between what is "pure" and what is "applied." Even the "purest" of mathematics may suddenly find "application." A problem of great importance to an electrical industry, which had failed of solution by its own engineers, has been solved by using methods of set-theoretic topology. Topics in matrix theory, topology, and set theory have been applied to production and distribution problems; abstract concepts of modern algebra find application in electronics; and mathematical logic is applied to the theory of automata and computing machines.

During the time of the Greeks, mathematics was considered to be, on the one hand, an attempt to describe the forms, quantitative and geometric, that one finds in the environment and, on the other hand, a description of an ideal world of concepts existing

---

[2] From *The Rise of the New Physics,* by A. d'Abro, 2nd rev. ed., Dover Publications, Inc., New York, 1951. Reprinted through permission of the publisher. From our viewpoint, both classical mechanics and relativity are conceptual systems whose loci lie equally in the scientific component of culture. What d'Abro has stated is that the scientific position of classical mechanics has changed, its former position being occupied by the newer theory of relativity; in the above diagram, it has shifted into a darker area.

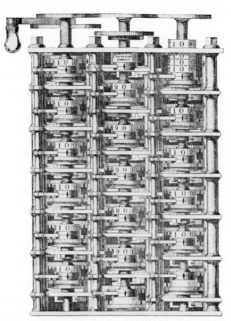

(*Top*) The computing element of Babbage's multiplier, invented by Charles Babbage in the 19th century. (The Bettman Archive.)

(*Bottom*) Transistorized circuits similar to the above are used in modern computers. (Eliot Elisofon, LIFE Magazine © Time, Inc.)

over and above the so-called real world. Following the abstractions of the nineteenth-century algebraists and the introduction of the non-Euclidean geometries during the early part of the nineteenth century, this dual character of mathematics was changed. Although one could consider that mathematics offered conceptual frames that one could more or less successfully fit to natural and cultural phenomena, the concepts were no longer embodiments of an independently existing realm of ideas, having an existence before and after the fact of their discovery, but only of a world of concepts continually under construction and having no existence until conceived in the minds of the mathematicians who create them. The locus of this conceptual world could now be precisely located, namely in culture itself (*cf.* White, 1949, Chapter X; also Wilder, 1950, 1953, 1960). Mathematics derives its concepts initially from the existing world of reality and uses them as a way of dealing with this reality; only now this "reality" embaces not just the physical environment, but the cultural—which includes the conceptual—environment. Concepts are just as real as guns or butter; merely ask the doubter how he could fight a war without them! The main difference between the applied mathematician and the pure mathematician is that they are dealing with different aspects of reality.

And this brings us to the matter of *freedom* in mathematics. Following the nineteenth-century developments, the mathematical world came to feel that it was no longer restrained by the world of reality, but that it could create mathematical concepts without the restrictions that might be imposed by either the world of experience or an ideal world to whose nature one was committed to limit his discoveries. One is reminded of the mathematician who, disgusted by the uses to which a backward and laggard world was putting scientific concepts, exclaimed, "Thank God that there is no danger of *my* work ever being put to practical use." He was giving expression to that kind of "freedom" that the mathematical world came to feel during the past century. The gentleman in question was probably not well aware of the nature of modern mathematics, or he would not have been so confident in his exhilaration. No one can escape his environment, and in particular no mathematician can escape his cultural environment. Even though he may believe that mathematics is not a science but an *art* and even though his motives be certainly artistic, whatever mathematics he creates cannot be other than conditioned by the mathematical milieu in which he has been trained; in short, his freedom is limited by the existing state of mathematics in his culture. His success as a mathematician will be measured by the quality of his contributions to the solutions of those problems that are outstanding at the time when he is active. No one can deny that, as an individual, he has the freedom to indulge in whatever mathematical fantasies he prefers; but if they happen not to have significance for the conceptual state of mathematics at the time, they will receive no recognition (unless, of course, he is "before his time" and constitutes therefore a "singularity"—see Chapter 4, Section 4.1).

Freedom in mathematics, like all "freedoms," is limited by the culture in which it is exercised. And so long as the "pure" mathematician selects his area of research from currently significant domains of mathematics, he may be assured that his accomplishments will be meaningful. For he, too, is an "applied mathematician," only the applications in his case are to the conceptual mathematical part of his culture—the dark area of the diagram. Moreover, his works will sooner or later almost inevitably find, directly or indirectly, "application" to nonmathematical aspects of culture—sometimes in the most unexpected places!

2 THE "FOUNDATIONS" OF MATHEMATICS   It appears to be a universal phenomenon in the evolution of culture, that when a culture has evolved sufficiently to achieve a certain degree of maturity, there then arises a need among its participants for an "explanation" of its origin. The classical case in ethnology is that of a group of people who in ancient times migrated from some parent group and evolved a culture of their own. The relation to the parent group becomes forgotten, because of the great lapse of time since the migration, as well as because of the absence of written records, and with the increased consciousness of its own identity, the bearers of the new culture find a need to bolster this identity with a narrative supplying an account of its genesis. Although fictional, the narrative acquires a sacred character that increases its meaningfulness and supplies the stability and feeling of security requisite for the continuance of the culture. For example, one can imagine the situation of the Mesa Verde Indians whose world view was probably limited to their immediate environment and who were probably at the mercy of both natural hazards and marauding bands of strangers. Such a group inevitably seeks solace and protection in a philosophy that justifies and enhances its culture; this philosophy encompasses both an "explanation" of the culture's origin and the basis for a ritualistic religion that serves as a source of protection from natural and supernatural dangers as well as a means of cementing the cultural ties of the tribe.

2.1 *The mathematical subculture*   An analogous phenomenon can be observed in the case of subcultures. And the mathematical subculture of modern Western culture[3] furnishes no exception. An interesting contrast between mathematics and the other sciences needs to be brought out at this point. Although physics and chemistry, for example, went through mystical periods during their development, the effects of these have largely worn off. It is generally recognized that these sciences are limited to an explanation of natural phenomena, and if the increased efficiency of means of measurement reveals that a physical theory is not truly representative of the situation that it was devised to explain, a suitable substitute theory is sought. And no modern scientist regards such an event as a threat to the security of science.

The situation in mathematics, on the other hand, has traditionally been quite different. Elements of mysticism survived for

---

[3] In "Western culture" are included those parts of non-Western cultures that have been acquired from the Western culture.

a long time in the mathematical community, and today Platonic idealism is apparently not uncommon. The faith in the "truth" of mathematical theories that has been sustained in the general culture is shared to a considerable extent by the mathematical subculture. The abstract nature of modern mathematics contributes to this view, despite the fact that most mathematicians of prominence concur in the doctrine that modern algebraic and geometric theories are true only in the sense that they are logical consequences of the axioms that form their bases. No mathematician who is familiar with the modern situation in mathematics will argue for the "truth" of either Euclidean or non-Euclidean geometry, for example. But in the case of those parts of mathematics that depend on the natural number system and its extensions, as well as on logical derivation therefrom—and this ultimately includes a good part of mathematics—there are those who argue for the absolute character of their conclusions.

It is not, necessarily, that there exists greater cohesion among the members of the mathematical community than in the community of physicists, but that the very nature of the field renders any threat to its conclusions of greater importance to mathematics than does the collapse of a physical theory to the physicists.

2.2 *The emergence of contradictions*   Attention has already been directed toward two of the threats to the soundness of mathematics: (1) the discovery of incommensurables and the Zeno paradoxes during the Greek era and (2) the realization of the inadequacy of the pre-nineteenth-century conception of the real number continuum (Chapter 3). The former was resolved by the Eudoxian theory of proportion (based on magnitudes), and the latter by the work of Weierstrass, Dedekind, and other nineteenth-century analysts who gave a seemingly conclusive definition of the real number continuum. But it has turned out to be impossible to analyze the real number continuum without bringing in the new notions of *set theory*.

The early attitude toward set theory was much like that which was in vogue regarding logic; indeed, many have treated it as a part of logic. This attitude was one of an unquestioning nature regarding the reliability of both logic and set theory. As we have observed, the Greeks brought the notion of proof by logic into mathematics—the Babylonians and Egyptians had no notion of such methods as *reductio ad absurdum,* for example. The trustworthiness of the Aristotelian laws of contradiction[4] and of the excluded middle[5] was not questioned, and mathematical conclusions reached by the use of such "laws" were considered absolutely reliable if the premises were.

In much the same way, the nineteenth-century mathematicians introduced set theory into mathematics; like logic, it was derived from experience with the finite collections of the physical and cultural environments. That extension of the classical logic and of set theory to infinite domains might lead to difficulties was not generally anticipated until around 1900, when a number of contradictions were found. One of the best known of these was given by B. Russell and may be described as follows: Let us call a set that is not an element of itself an *ordinary* set. All the sets of our everyday experience are ordinary; for example, the set of all players on a basketball team is not itself a player, nor the set of all books in a library a book. (One has to exercise some ingenuity to find a set that is not ordinary; a common suggestion is the set of all abstract ideas—which is certainly itself an abstract idea and consequently an element of itself.)

Trouble enters when one considers the set, S, of all ordinary sets. By the logical law of the excluded middle, the set S either is ordinary or is not ordinary. However, if S is ordinary, then S is *not* an element of itself by definition. But this can be the case only if S is not ordinary, since all ordinary sets are elements of S. On the other hand, if S is not ordinary, then S has itself as an element; but elements of S are all ordinary, so S must be ordinary.

---

[4] Roughly, a meaningful statement cannot be both true and false.
[5] If S is a meaningful statement, then either S is true or it is false.

To summarize, if S is ordinary then it is not ordinary; and if S is not ordinary then it is ordinary!

Thus the security of mathematics was again threatened after the self-satisfaction produced by the "safe" founding of analysis in the nineteenth century; a "crisis" somewhat like that of the Greek era was again encountered. A new foundation for the whole of mathematics seemed necessary to meet this crisis, not just a revised formulation of the real number continuum; for all parts of mathematics depended to greater or lesser extent on logic and set theory. Some of the most capable of early-twentieth-century mathematicians undertook to set matters right. Best known of these attempts were those of the English mathematicians and philosophers B. Russell and A. N. Whitehead, whose work was presented in their monumental *Principia Mathematica;* the eminent German mathematician D. Hilbert; and the Dutch mathematician L. E. J. Brouwer.

Throughout the nineteenth century, the urge for an "explanation" of the true nature of mathematics had provided a hereditary stress that resulted in a number of new foundations not only for particular fields such as geometry, but for the whole of mathematics. In the latter respect, of particular importance were the works of Frege and Peano. Frege, a German mathematician, insisted that number and all of mathematics can be grounded in logic —a doctrine sometimes called "the logistic thesis." The Italian mathematician Peano and his disciples refined and utilized the axiomatic method to achieve a basis for mathematics, introducing therewith a symbolism that allowed greater precision of statement than the language of ordinary discourse, such as was used in Euclid's geometry, for example. It was largely under the influence of the works of Frege and Peano that the work of Russell and Whitehead was fashioned. In the *Principia Mathematica* an attempt was made to derive mathematics from self-evident universal ("tautological") logical truths. But in retrospect it becomes obvious that as the work proceeded to the higher realms of mathematical abstraction, it became necessary to introduce axioms that could hardly be admitted as constituting "self-evident logical truths." Means for avoiding contradiction were, however, worked out.

Hilbert's approach was more frankly axiomatic, in that his basic terms and propositions, though dressed up in a symbolism similar to that of *Principia Mathematica,* were not presented as constituting any more than a set of basic assumptions, neither true nor false of themselves, but "formulas" from which one would hope to derive, by carefully formulated "finitistic" methods, the whole of mathematics free of contradiction. This type of work, called "formalism," was ultimately presented in a two-volume work entitled *Grundlagen der Mathematik* ("Foundations of Mathematics") in which Hilbert's eminent student P. Bernays collaborated.

Of a radically different character—so much so as to form virtually a "cultural singularity" (*cf.* Chapter 4, Section 4.1)— was the doctrine of the nineteenth-century mathematician Leopold Kronecker. His "explanation" of the nature of mathematics consisted of the assertion that it was a construction based on the natural numbers, which, in turn, were an outgrowth from man's "intuition." Unlike his contemporaries, who were following the course of mathematical evolution unhampered by most of the mystical restrictions regarding the "reality" of negative or irrational numbers that beset their forebears, Kronecker avoided all use of numbers that could not be constructed (as can, for instance, fractions like ⅔) from natural numbers. Virtually no one agreed with him; he asserted that numbers like pi, for example, simply do not "exist," since there are apparently no ways of constructing them from natural numbers.

After the crises of the turn of the century, Kronecker's thesis was reaffirmed (in modified form) by Brouwer, and in a series of profound papers this young and brilliant Dutch mathematician formulated the type of mathematics that came to be known as intuitionism. The logic that had been introduced into mathematics by the Greeks was tossed overboard, except for what could be salvaged through use of the constructive methods of intuitionism. In particular, use of the law of the excluded middle, so important in *reductio ad absurdum* proofs, was no longer permissible except for finite sets. For example, of any *finite* set of natural numbers, it was permissible to assert that either at least one of the numbers was even or none was even. There exists an elementary constructive way of demonstrating such use of the law of the excluded middle, namely by examining the numbers one by one! But the same assertion about an arbitrary *infinite* set of natural numbers could not be made—unless, of course, one could in some way point to an even number in the set, which would be an admissible constructive act.

The great advantage of the intuitionistic philosophy was its freedom from contradiction—limitation to constructive methods guaranteed this. But its fatal defect was that it could not derive, using only its constructive methods, a major portion of the concepts that were regarded as being among the greatest mathematical achievements of the modern era. Viewed from the standpoint of the present day, intuitionism can be regarded as an attempt to stem the flow of mathematical evolution—a kind of cultural resistance. The stresses (hereditary) formed by the desire for an "explanation" of mathematics and the demand for protection against the threat of contradiction were clearly forcing the mathematical world to take action; but not such drastic action as intuitionism demanded. The latter would be like expecting a primitive tribe to kill off most of its members to avoid their possible annihilation by a threatening enemy.

Despite such considerations, intuitionism had a great and seemingly beneficial influence. A number of prominent mathematicians shared in some, or all, of its tenets—for example, H. Poincaré and H. Weyl. But, and of more importance, its doctrine of constructivity was found to be adaptable to numerous situations within the framework of conventional mathematical theory.

2.3 *Mathematical logic and set theory*   Along with these developments, and flowing out of them to a considerable extent, there evolved a searching analysis of both logic and set theory—the two most "taken for granted" features of mathematical *method*. As might be expected, it is only with considerable difficulty that one can disentangle the complex of evolutionary forces that were operative during this period. On the one hand, the hereditary stress caused by the desire for an "explanation" of the nature of mathematics—intensified by the crisis caused by the discovery of the contradictions—induced the researches culminating in the three "schools of thought" represented by logicism, formalism, and intuitionism. On the other hand, there was great cultural resistance to such investigations in the mathematical community; many mathematicians would take no part in them, assuming a scornful or disdainful attitude. Evident also was cultural lag, in that many —perhaps the majority—took no interest in the situation (perhaps a continuation of the old attitude that the important thing is to *make* mathematics and not worry about the consequences).

Of greatest interest, however, is the manner in which both logicism and formalism utilized *symbolization*. The trend to more intensive use of ideograms, evidenced in both the calculus of Leibniz and Newton and in the various algebras of the eighteenth and nineteenth centuries, can be considered to have reached a zenith in the doctrines of both logicism and formalism—later to merge in modern mathematical logic. It became clear, with the passage of time and the advantage of hindsight, that both systems —logicism and formalism—were essentially attempts to base mathematics on carefully selected sets of axioms expressed in purely ideographic symbols, with directions for the methods by which new formulas ("theorems") could be derived ("proved") therefrom. If man is to be distinguished from other animals by his ability to symbolize, then certainly here was a most human activity!

In 1931, however, hopes for the success of such programs as those of Russell-Whitehead and Hilbert were dashed by the demonstration, by the young Austrian mathematician K. Gödel, of the impossibility of either achieving a complete description of mathematics thereby or proving the consistency of such systems within their own framework. Earlier, the Scandinavian logician Skolem had initiated work that led, ultimately, to the conclusion that no complete foundation could ever be achieved for the theory of sets. Moreover, it soon became apparent that neither logic nor the theory of sets, when analyzed by the powerful methods that had been developed in modern mathematical logic, could be characterized as a unique theory; instead, it turned out to be possible to develop a variety of logics and set theories. Thus the forces of generalization and diversification invaded what had been considered the most absolute areas of human thought. Even the natural numbers, the most ancient and inviolable of mathematical entities, defied definitive definition!

This may be considered a partial victory for intuitionism, whose doctrine of the natural numbers as the foundation of mathematics seemed now to have found support. And the mathematician who had refused to participate in the attempt to give a consistent and complete foundation for modern mathematics could now feel that he had been justified in accepting the classical methods of logic that he had inherited, as well as those elementary portions of set theory that were all he usually needed for his work. But in doing so he had to admit that he was accepting an intuitive foundation that had been bequeathed him by his culture. For the logic and the set theory that he uses have been a product of evolution just as much as have his number systems, geometries, and other theories.

Mathematics is thereby put in a perspective which reveals that its position, as a science, is not unlike that of other sciences. The major difference between mathematics and the other sciences, natural and social, is that whereas the latter are directly restricted in their purview by environmental phenomena of a physical or social nature, mathematics is subject only indirectly to such limitations. As we have seen, mathematics, almost from its inception, has been becoming increasingly self-sufficient. The problems on which the modern mathematician works arise mainly either from theories already existing within mathematics or from the theories of the sister sciences and are consequently entirely of cultural origin. That his most powerful symbolic tools and his powers of abstraction and generalization have failed the mathematician insofar as "explaining" what mathematics is, or in providing a secure "foundation" and absolutely rigorous methods, is of no more consequence than the failure of other sciences to achieve final and accurate explanations of the phenomena that they study. If certain methods lead to contradiction, they must be modified, just as a physical scientist must frequently made modifications. Perfect rigor and absolute freedom from contradiction in mathematics are no more to be expected than are final and exact explanations of natural or social phenomena. The present-day situation in mathematics is made all the more interesting by the realization that there cannot, because of the cultural nature of mathematics, ever be an end to its evolution. So long as the progress of man's cultural evolution continues uninterrupted, just so long will mathematics —like physics, chemistry, biology, and the social sciences—continue to evolve more abstract, scientifically effective, and marvelous concepts.

3 MATHEMATICAL EXISTENCE   Among the problems that have been touched upon here and there is that of *mathematical existence*. In what sense, specifically, do all these abstract concepts of number, geometry, and set theory that have evolved, exist? This is a problem that has engaged philosophical discussion ever since the ancient Greeks. As we have seen, the Pythagoreans gave to number—the "natural number"—an absolute status that placed it above and beyond human meddling. Plato conceived of an ideal universe in which resided perfect models of all the geometric configurations known to antiquity. The assumption made in the present work is that the only reality mathematical concepts have is as cultural elements or artifacts. The advantage of this point of view is that it permits one to study the manner in which mathematical concepts, as cultural elements, have evolved and to offer some explanation of why and how concepts are created from the syntheses produced by cultural forces in the minds of individual mathematicians. Moreover, the mysticism that creeps surreptitiously into most forms of idealistic attitudes toward mathematical existence disappears. In addition, misunderstanding and confusion regarding the *permissibility* of certain concepts, hangovers from the influence of environmental stresses of a physical nature on mathematics, are cleared up. For example, an infinite decimal is not something that "just goes on and on without end." It is to be conceived as a *completed* infinite, just as one conceives of the totality of natural numbers as a completed infinity. Symbolically, it may be considered a *second-order* symbolism, in that it is not susceptible to complete perception, but is only *conceptually perceivable*.

Because of the origin of numbers, as well as of geometry, in the world of physical reality, both philosophers and mathematicians have repeatedly sought to justify the "reality" of mathematical concepts by appeal to physical reality. Thousands of pages have been devoted to discussions of the question whether Euclidean geometry is "true" or not; in particular, is the "time continuum" represented faithfully by the Euclidean straight line? From a cultural standpoint, such questions are meaningless. The concept of number is an existing *cultural* entity, whose origin and evolution were induced by cultural stress of environmental and hereditary character. And the concept of an infinite totality of natural numbers is not, as seems to be often asserted, open to argument as regards its *existential* nature. A finite mathematics based on the concept of natural number alone may be sufficient in a culture that has advanced only to the stage where such numbers are adequate for scientific purposes. But such theories as are embodied in the calculus and, more generally real analysis, which were themselves largely the product of environmental stresses exerted by mechanics, physics, and the like, ultimately created hereditary stresses that demanded an infinite mathematics for their further development. Whether infinite totalities exist in the physical world has nothing to do with the case. What matters is, do the concepts lead to fruitful mathematical developments? And the answer is that they do and, moreover, they solved the crisis that faced analysis during the last three centuries. Of course, they have brought on new crises, such as are embodied in the theory of sets —but these are in their turn being subjected to searches for their solution. D'Alembert's advice to "go forward and faith will come to you" is excellent until the mathematical edifice threatens to collapse; then the courage to launch forth into a new conceptual world is needed to save the day.

4 "LAWS" GOVERNING THE EVOLUTION OF MATHEMATICAL CONCEPTS   The forces listed in Chapter 4, Section 4, as operative in the evolution of mathematical concepts should be studied further, particularly in regard to the manner in which they manifest themselves and in regard to their completeness. Are there any characteristic principles or "laws" that they seem to obey, for instance? More case histories need to be analyzed, but this would necessitate going into technicalities of a mathematical nature not contemplated in the present work.[6]

By way of conclusion, however, certain principles are suggested below, which appear to be worthy of study with a view to their justification or refutation:

1. At any given time, there will evolve only concepts that are so related to the existing mathematical culture as to increase its utility in meeting the demands of either its own hereditary stresses or the environmental stresses from the host culture.

2. The admissibility and acceptance of a concept will be de-

---

[6] One of the author's students has made a case study of a portion of mathematical logic; see Judith Ann Orman Lewis, "The Evolution of the Logistic Thesis in Mathematical Logic," Doctoral Dissertation, University of Michigan, 1966.

cided by its degree of fruitfulness. In particular, a concept will not be forever rejected because of its origin or on the grounds of metaphysical criteria such as "unreality."

3. The extent to which a concept continues to be mathematically important is dependent both on the symbolic mode in which it is expressed and on its relationship to other concepts. If one symbolic mode tends toward obscurity, or even outright rejection of the concept, then—assuming the usefulness of the concept—a more easily comprehensible form will evolve. If a group of concepts are so related as to make consolidation of them all within a more general concept feasible, then the latter will evolve.

4. If the advance of a mathematical theory will be promoted by the solution of a certain problem, the conceptual structure of the theory will evolve in a manner to permit the eventual solution of the problem. The circumstances are then likely to be such that the solution will be found (but not necessarily published) by several investigators working independently. (Proof of unsolvability is regarded as a "solution" of the problem, as evidenced by the history of squaring the circle, trisection of the angle, and the like.)

5. The opportunities for diffusion, such as may be provided by a universally accepted symbolism, increased outlets for publication, and other means of communication, will have a direct effect on the rate at which new concepts evolve.

6. Needs of the host culture, especially when accompanied by increased facilities that may be provided for the nourishment of the mathematical subculture, will result in the evolution of new conceptual devices to meet the needs.

7. A static cultural environment will eventually stifle the development of new mathematical concepts. A similar effect will result from an adverse political or general antiscientific atmosphere.

8. A crisis, such as may be produced by exposure of an inconsistency in the current conceptual structure, or the inadequacy thereof, will stimulate accelerated evolution of new concepts.

9. New concepts will usually depend on concepts that are only intuitively perceived at the time, but that will ultimately generate new crises due to their inadequacies. Similarly, solution of an outstanding problem will create new problems.

10. Mathematical evolution remains forever a continuously progressing affair, limited only by the contingencies described in items 5 through 7.

4.1 *Discussion* By "host culture" is meant the culture of which the mathematics forms a subculture. Unfortunately it is not uniquely definable. At certain times, historically speaking, it would have been determined by national boundaries, as in the case of ancient Chinese mathematics. In modern times, the host culture usually transcends national boundaries, unless political forces intervene.

In the case of 6, it is revealing to consider the state of mathematics in the United States since the inception of World War II. The needs, economic and political, of the host culture stimulated the development of computing machines and, eventually, a whole new chapter in theoretical and applied mathematics. Other new mathematical structures were also directly attributable to wartime needs. The subsequent governmental policy of fostering mathematical research by means of subsidies from the defense agencies and the National Science Foundation resulted in the accelerated evolution of new mathematical concepts as well as in an increase in the number of students who eventually became mathematicians.

Item 1 is fairly self-evident; new concepts are always related in some manner to the existing body of mathematical concepts, and their invention is instigated by the need for settling pressing problems presented either by the existing mathematical culture (hereditary stress) or by the host culture (environmental stress). But a person who might have had an admirable mental equipment for creating modern algebraic concepts would certainly never have done so if he happened to be a citizen of ancient Greece.

A good example in support of 2 is the ultimate acceptance of negative numbers into mathematics, as well as of "imaginary" numbers such as $\sqrt{-1}$. So long as these types of numbers were not indispensable, they were rejected on the grounds that they were "unreal" or "fictional."

Item 3 is well exemplified by the evolution of the place value system for representing real numbers. The Babylonian ciphers were replaced by other, simpler symbols; but the Babylonian place value system, extended to fractions, survived and forms the origin of our own numerical system. Moreover, its permanent use was eventually assured when it was found possible to symbolize, in a conceptual fashion, the arbitrary real number by means of the infinite decimal. The professional mathematician who has observed the development of mathematics over the past half century will have no difficulty in recalling instances, on a higher mathematical level, of the simplification of symbolic devices and the consolidation of related concepts within a more general conceptual framework.

The best-known instance illustrating item 4 is the classical problem concerning the Euclidean parallel axiom. The solution, usually attributed to Gauss, Lobachewski, and Bolyai, working independently during the first third of the nineteenth century, was clearly about to burst forth, as the reading of any account of the period shows (see, for example, Bell, 1945, pp. 325-326). As one approaches nearer to the present, he finds that instances abound.

It seems hardly necessary to elaborate on items 5 and 7. As might be expected, ancient Chinese mathematics, like the host culture itself, became static; the decline of Greek creativity in mathematics coincided with the general cultural decline of the period and may constitute another example.

In regard to item 8, one can point first to the crisis in Greek mathematics precipitated by the discovery of irrational magnitudes and the Zeno paradoxes, followed by the intense creative period during which Greek geometry was developed. The crisis regarding the foundations of analysis culminated in the concept of the real number system as related in Chapter 3. The latter also provides an excellent illustration of item 9, in that it introduced into mathematics the concept of "set," which in turn caused a crisis about the turn of the century when it was discovered that contradictions resulted from unbridled use of the concept and that it must in its turn be subjected to analysis.

The assertion of item 10 can be adequately supported only by a more complete treatment of the history of mathematics; it is certainly the case that creative professional mathematicians would generally agree to it on the basis of their own experiences. It is, indeed, virtually a corollary of items 8 and 9.

4.2 *Conclusion* In view of the fact that the power and utility of mathematics have increased as its conceptual patterns have become more and more abstract, it seems justifiable to formulate what might be termed the Magna Charta of the creative worker in the field:

There shall be established no limit to the "intrinsic" character or nature of conceptualization, other than what may be imposed by the scientific merit of its consequences. The judgment regarding scientific merit is to be *post facto*. In particular, a concept will not be rejected because of such vague criteria as "unreality" or because of the manner in which it has been devised.

Bibliography

Archibald, R. C.
 1949 "Outline of the History of Mathematics," *American Mathematical Monthly,* Vol. 56, Supplement; 6th ed. revised and enlarged.

Barnes, H. E.
 1965 *An Intellectual and Cultural History of the Western World,* 3d ed. revised, New York, Dover.

Bell, E. T.
 1931 *The Queen of the Sciences,* Baltimore, Williams and Wilkins.

 1933 *Numerology,* Baltimore, Williams and Wilkins.

 1937 *Men of Mathematics,* New York, Simon and Schuster (reprinted by Dover).

 1945 *The Development of Mathematics,* 2nd ed., New York, McGraw-Hill.

Bourbaki, N.
 1960 *Éléments d'Histoire des Mathématiques,* Paris, Hermann.

Boyer, C. B.
 1944 "Fundamental Steps in the Development of Numeration," *Isis,* Vol. 35, pp. 153-168.

 1949 *The History of the Calculus and Its Conceptual Development,* New York, Dover.

 1959 "Mathematical Inutility and the Advance of Science," *Science,* Vol. 130, pp. 22-25.

Bredvold, Louis I.
 1951 "The Invention of the Ethical Calculus," in *The Seventeenth Century: Studies in the History of English Thought and Literature from Bacon to Pope,* edited by R. F. Jones et al., Stanford, Calif., Stanford University Press.

Bridgman, P. W.
  1927 *The Logic of Modern Physics,* New York, Macmillan.
Chiera, E.
  1938 *They Wrote on Clay,* Chicago, University of Chicago Press, Phoenix Books.
Childe, V. G.
  1946 *What Happened in History?,* New York, Penguin Books, Pelican Book P6.
  1948 *Man Makes Himself,* London, Watts & Co., The Thinkers Library No. 87.
  1951 *Social Evolution,* New York, Henry Schuman.
Conant, L. L.
  1896 *The Number Concept,* New York, Macmillan.
Coolidge, J. L.
  1963 *A History of Geometrical Methods,* New York, Dover.
Courant, R., and H. Robbins
  1941 *What is Mathematics?,* New York, Oxford University Press.
D'Abro, A.
  1951 *The Rise of the New Physics,* 2nd ed., New York, Dover.
Dantzig, T.
  1954 *Number, the Language of Science,* 4th ed., New York, Macmillan.
Dubisch, R.
  1952 *The Nature of Number,* New York, Ronald Press.
Eves, H.
  1953 *An Introduction to the History of Mathematics,* New York, Rinehart and Co.
Firestone, F. A.
  1939 *Vibration and Sound,* 2nd ed.
Frege, G.
  1884 *Die Grundlagen der Arithmetik,* Breslau, Wilhelm Koelner.
  1950 *The Foundations of Arithmetic,* Oxford, Basil Blackwell.
Freudenthal, H.
  1946 *5000 Years of International Science,* Groningen, Noordhoff.
Gandz, S.
  1948 "Studies in Babylonian Mathematics," *Osiris,* Vol. 8, pp. 12-40.
Hadamard, J.
  1949 *The Psychology of Invention in the Mathematical Field,* Princeton, N.J., Princeton University Press.

Hardy, G. H.
1941 *A Mathematician's Apology,* Cambridge, England, The University Press.

Heath, T. L.
1921 *A History of Greek Mathematics,* 2 vols., Oxford, England, Oxford University Press.
1926 *The Thirteen Books of Euclid's Elements,* 3 vols., 2nd rev. ed., Cambridge, England, Cambridge University Press.

Hopper, V. F.
1938 *Medieval Number Symbolism,* New York, Columbia University Press.

Huxley, J.
1957 *Knowledge, Morality and Destiny,* New York, New American Library of World Literature, Mentor Book.

Karpinski, L. C.
1925 *The History of Arithmetic,* New York, Rand McNally.

Kasner, E., and J. R. Newman
1940 *Mathematics and the Imagination,* New York, Simon and Schuster.

Klein, F.
1892 "A Comparative Review of Recent Researches in Geometry," *Bulletin of the New York Mathematical Society,* Vol. 2 (1892-3), pp. 215-249; translated by M. W. Haskell.
1932 *Elementary Mathematics from an Advanced Standpoint,* translated by E. R. Hedrick and C. A. Noble, Part I, New York, Macmillan.
1939 *Elementary Mathematics from an Advanced Standpoint,* Part II, *Geometry,* 3rd ed., New York, Macmillan.

Kline, M.
1953 *Mathematics in Western Culture,* New York, Oxford University Press.

Kroeber, A. L.
1948 *Anthropology,* rev. ed., New York, Harcourt, Brace & World.

Kuhn, T. S.
1962 *The Structure of Scientific Revolutions,* Chicago, University of Chicago Press.

Malinowski, B.
1945 *The Dynamics of Culture Change,* New Haven, Conn., Yale University Press.

Menninger, K.
1957 *Zahlwort und Ziffer,* Vol. 1, Göttingen, Vandenhaeck and Ruprecht.

Merton, Robert K.
   1957 "Priorities in Scientific Discovery: A Chapter in the Sociology of Science," *American Sociological Review,* Vol. 22, pp. 635-659.
   1961 "Singletons and Multiples in Scientific Discovery: A Chapter in the Sociology of Science," *Proceedings of the American Philosophical Society,* Vol. 105, pp. 470-486.
Moritz, R. E.
   1914 On Mathematics and Mathematicians, New York, Dover.
Nagel, E., and J. R. Newman.
   1958 *Gödel's Proof,* New York, New York University Press.
National Council of Teachers of Mathematics
   1957 *Insights into Modern Mathematics,* Twenty-third Yearbook, Washington, D.C.
Neugebauer, O.
   1957 *The Exact Sciences in Antiquity,* 2nd ed., Providence, R.I., Brown University Press.
   1960 "History of Mathematics, Ancient and Medieval," Chicago, Ill., *Encyclopaedia Britannica,* Vol. 15, pp. 83-86.
Poincaré, H.
   1946 *The Foundations of Science,* translated by G. B. Halstead, Lancaster, Pa., Science Press.
Rosenthal, A.
   1951 "The History of Calculus," *American Mathematical Monthly,* Vol. 58, pp. 75-86.
Russell, B.
   1937 *The Principles of Mathematics,* 2nd ed., New York, W. W. Norton.
Sahlins, M. D., and E. R. Service, editors
   1960 *Evolution and Culture,* Ann Arbor, Mich., University of Michigan Press.
Sánchez, G. I.
   1961 *Arithmetic in Maya,* Austin, Texas, published by author (2201 Scenic Drive).
Sarton, G.
   1935 "The First Explanation of Decimal Fractions and Measures (1585), Together with a History of the Decimal Idea and a Facsimile (no. xvii) of Stevin's Disme," *Isis,* Vol. 23, pp. 153-244.
   1952 "Science and Morality," in *Moral Principles of Action,* edited by Ruth N. Anshen, New York, Harper & Row, p. 444.
   1959 *A History of Science,* 2 vols., Cambridge, Mass., Harvard University Press.

Seidenberg, A.
    1960 "The Diffusion of Counting Practices," *University of California Publications in Mathematics,* Vol. 3, No. 4, pp. 215-300.

Smeltzer, D.
    1953 *Man and Number,* London, Adam and Chas. Black.

Smith, D. E.
    1923 *History of Mathematics,* 2 vols., Boston, Houghton-Mifflin.

Struik, D. J.
    1948a *A Concise History of Mathematics,* 2 vols., New York, Dover.
    1948b "On the Sociology of Mathematics," in *Mathematics, Our Great Heritage,* edited by W. L. Schaaf, New York, Harper, pp. 82-96.

Szabo, A.
    1960 "Anfang des Euklidischen Axiomsystems," *Archive for History of Exact Sciences,* Vol. 1, pp. 37-106.

Thureau-Dangin, F.
    1939 "Sketch of a History of the Sexagesimal System," *Osiris,* Vol. 7, pp. 95-141.

Tingley, E. M.
    1934 "Calculate by Eights, Not by Tens," *School Science and Mathematics,* Vol. 34, pp. 395-399.

Tylor, E. B.
    1958 *Primitive Culture,* 2 vols., New York, Harper Torchbooks 33, 34.

Van der Waerden, B. L.
    1961 *Science Awakening,* translated by Arnold Dresden, New York, Oxford University Press.

Waismann, F.
    1951 *Introduction to Mathematical Thinking,* translated by T. J. Benac, New York, Frederick Ungar.

Weyl, H.
    1949 *Philosophy of Mathematics and Natural Science,* Princeton, N.J., Princeton University Press.

White, L. A.
    1949 *The Science of Culture,* New York, Farrar, Straus; also published in paperback edition by Grove Press, as Evergreen Book E-105.
    1959 *The Evolution of Culture,* New York, McGraw-Hill.

Wilder, R. L.
    1950 "The Cultural Basis of Mathematics," *Proceedings of the International Congress of Mathematicians,* pp. 258-271.

1953 "The Origin and Growth of Mathematical Concepts," *Bulletin of the American Mathematical Society,* Vol. 59, pp. 423-448.

1960 "Mathematics: A Cultural Phenomenon," in *Essays in the Science of Culture,* edited by G. E. Dole and R. L. Carneiro, New York, T. Y. Crowell, pp. 471-485.

1965 *Introduction to the Foundations of Mathematics,* 2nd ed., New York, John Wiley and Sons.

Index